Chaotic Signal Processing

Chaotic Signal Processing

Edited by

Henry Leung

University of Calgary
Calgary, Alberta
Canada

siam.
Society for Industrial and Applied Mathematics
Philadelphia

高等教育出版社
HIGHER EDUCATION PRESS
Beijing, People's Republic of China

10 9 8 7 6 5 4 3 2 1

Library of Congress Cataloging-in-Publication Data

Chaotic signal processing / edited by Henry Leung, University of Calgary, Calgary, Alberta, Canada.
pages cm – (Applied mathematics)
Includes bibliographical references and index.
ISBN 978-1-61197-325-9
1. Signal processing–Mathematics. 2. Chaotic behavior in systems. I. Leung, Henry, editor of compilation.
TK5102.9.C47145 2014
621.382'20151539–dc23
2013037734

List of Contributors

Jonathan N. Blakely
U. S. Army Research, Development and Engineering Command, Redstone Arsenal, Alabama 35898, USA

Thomas L. Carroll
US Naval Research Laboratory, Code 6362, Washington, DC 20375, USA

Ned J. Corron
U. S. Army Research, Development and Engineering Command, Redstone Arsenal, Alabama 35898, USA

William L. Ditto
Department of Physics and Astronomy, University of Hawaii at Manoa, Honolulu, HI 96822, USA

Benjamin C. Flores
Department of Electrical and Computer Engineering, University of Texas at El Paso, 500 University Ave., El Paso, TX 79928, USA

Ajeesh Kurian
Novatel Inc., 1120 - 68th Avenue NE, Calgary, Alberta, T2E 8S5, Canada

Ying-Cheng Lai
School of Electrical, Computer and Energy Engineering, Arizona State University, Tempe, AZ 85287, USA

Henry Leung
Department of Electrical and Computer Engineering, University of Calgary, 2500 University Drive NW, Calgary, Alberta, T2N 1N4, Canada

Abraham Miliotis
Department of Biomedical Engineering, University of Florida, Gainesville, FL 326611-6131, USA

Krishnamurthy Murali
Department of Physics, Anna University, Chennai 600 025, India

Shawn D. Pethel
U. S. Army Research, Development and Engineering Command, Redstone Arsenal, Alabama 35898, USA

Frederic J. Rachford
US Naval Research Laboratory, Code 6362, Washington, DC 20375, USA

Sudeshna Sinha
Indian Institute of Science Education and Research (IISER) Mohali, SAS Nagar, Sector 81, Mohali 140 306, India

Berenice Verdin
Department of Electrical and Computer Engineering, University of Texas at El Paso, 500 University Ave., El Paso, TX 79928, USA

Contents

Preface

Chaos theory is considered in signal processing because it may provide an alternative to statistical signal processing. Since chaos is inherently a deterministic dynamical behavior, it has the potential to provide a more fundamental description of a signal process when compared with use of statistics. In addition, chaos can be generated by simple nonlinear systems and may reduce the complexity of signal processing in many applications. There have been various applications of chaotic signal processing in different fields, such as biomedical signal processing (for example, electroencephalography), acoustic signals, speech, radar, and communications.

This book starts with an overview of chaotic signal processing and then presents current research results of contributors on different applications. Key contents include application of nonlinear dynamics to radar target recognition, an exactly solvable chaos approach for communications, a chaotic approach for reconfigurable computing, system identification using chaos, design of a high resolution LADAR system based on chaos, and the use of chaos in compressive sensing.

Henry Leung
Editor

Chapter 1

An Overview of Chaotic Signal Processing

Henry Leung

Abstract: Chaos theory has been applied to signal processing for the past two decades. This chapter provides a brief summary of some popular signal processing applications in which chaos has been considered, including biomedical signals, acoustic, speech, radar signals, and communications. We describe the signal processing techniques that have been developed to process chaotic signals. In particular, we consider detection and estimation techniques for chaotic signals, including detection and estimation of chaotic signals in noise and detection and estimation of signals in chaotic noise background. This chapter concludes with a brief introduction of current chaotic signal processing developments presented in this book.

1.1 ▪ Introduction

Chaos theory has been considered in signal processing for the past two decades. As many signals possess random appearance, statistics has been a major tool in performing signal analysis. Many researchers find chaos theory interesting as it may provide an alternate to statistical signal processing. There are many applications of chaos in different fields of signal processing, including biomedical signal processing such as electroencephalography (EEG) [1, 2], acoustic signals [3, 4], radar [5, 6], communications [7, 8], and security [9, 10]. A natural approach is to investigate whether chaotic phenomenon exists in these signals, in other words, whether chaos can provide good models for these signal processes.

In the biomedical application, epileptic seizures is one of the first being investigated using nonlinear dynamics. Epileptic seizures reflect a pathological state of the brain activity. During the seizure, the EEG activity may suddenly switch into oscillating modes. Researchers have shown that chaos can be used to describe EEG signals [1, 11]. EEG signals can be decomposed into five EEG subbands: delta (0–4 Hz), theta (4–8 Hz), alpha (8–12 Hz), beta (13–20 Hz), and gamma (30–60 Hz). EEG is usually assumed to represent the dynamics of the entire brain as a unified system. It is demonstrated using correlation dimension estimation that the dynamics of brain activity could be reconstructed. A relatively large difference is observed between the dimensionalities of the brain attractors and the epileptic states. A system of differential equations is further developed in [12], based on anatomical and physiological properties of the olfactory system. This model is shown to generate chaotic activity resembling the basal EEG for some parameter ranges and can simulate the activity observed in epileptic seizures of the olfactory systems.

Since different states of brain activity are shown to have different dynamic invariant measures, such as correlation dimensions and Lyapunov exponents [13], chaotic EEG analysis is brought to clinical application by using it to characterize Alzheimer's disease. It is shown in [14] that patients with Alzheimer's disease have a significantly lower correlation dimension and the first positive Lyapunov exponent values when compared to those for healthy people of similar ages. In other words, Alzheimer's disease has a less chaotic dynamical behavior than a healthy brain.

EEG signals also exhibit nonstationary behaviors, and a wavelet-chaos approach [15] is applied to characterize subband signals in delta, theta, alpha, and beta to identify potential markets of abnormality in Alzheimer's disease. It is shown in [2] that changes in brain dynamics are localized to certain frequency bands and electrode loci and new potential markers of abnormality are discovered based on the chaotic analysis.

Besides EEG signals, other biomedical signals such as electrocardiography (ECG) [16] and infant breathing signals [17] have also been investigated using chaotic analysis. It is shown that nonlinear dynamic provides a useful tool to quantify changes in healthy cardiovascular systems related to age [16]. Dynamical invariants such as the correlation dimensions and largest Lyapunov exponents are further used to detect ECG arrhythmia and to classify different arrhythmia types [18].

For acoustic signals, acoustic chaos is found in the sound output of water when ultrasound waves are focused into the liquid. The acoustic cavitation noise is shown to have fractal correlation dimension and positive Lyapunov exponents [19]. A resonator is further placed on three piezoelectric transducers in which one is a transmitter and two are receivers. High resolution acoustic spectroscopy is used to study the transitions between regularity and chaos [3]. Acoustic chaos is also found in sonic infrared imaging [4, 20], in which a heating effect is combined with acoustic sound to create an efficient nondestructive crack detection system. The transition from fundamental harmonic to chaotic behavior is shown to be important in obtaining a high-quality sonic infrared image.

Another application of chaos in acoustics is the modeling of underwater acoustic waveguides using ray tracing. Chaos is first applied to ray tracing to predict a radar cross section of concave structures such as ducts [21]. The Lyapunov exponent is found to be related to the accuracy of the geometry and hence the prediction. In [22], an approximate analytical approach is developed to analyze the chaotic behavior of rays in a deep-sea acoustic waveguide. This approach can be used to investigate the sound travel time between the source and the receiver. The underwater propagation effect is further investigated in [23]. It is shown in [23] that the exponential increment of local instability has an irregular pattern for chaotic rays, while it is either zero or has a regular behavior for regular rays. The launching basin is then introduced to identify regions of desirable ray launching parameters. In addition to analyzing the propagation effects, chaos has also been applied directly in sonar systems [24]. The chaotic pulse position modulation is used in [24] to enhance the ranging performance of sonar sensors for robotic navigation. Chua's circuit is used to implement the chaotic pulse position modulation, and this low cost approach is shown to improve the accuracy of sonar measurement experimentally. In fact, chaos modulation is also considered for underwater communications [25], in which a chaotic direct sequence spread spectrum system is developed. With a dual unscented Kalman filter demodulator, the chaos communication system is shown to achieve the desired low probability of interception. The results are verified in an experimental sea trial [26]. In addition to propagation and modulation, acoustic engineers have also applied chaos to acoustic emission signals. Ultrasonic signals are used to detect cracking and microdeformation. Correlation dimension is found to be a reliable feature for acoustic emission monitoring [27].

Another main application of chaos to acoustics is speech processing. Using spectrogram, bifurcation and chaotic episodes are found in cries of infants [28]. It is also observed that chaos occurs less frequently with increasing age of children, and this interesting observation could be of diagnostic relevance. For adult speech signals, it is also reported in [29] that measured speech signals have correlation dimensions in the range from 2 to 4, and the corresponding nonlinear predictor code has a 3dB prediction gain. In [30], experiments are conducted to illustrate that bifurcation, parameter control and attractors can be used to understand the underlying dynamics of speech categorization. In particular, a dynamical model is developed and two experiments are performed to validate the dynamical nature of speeches by observing perceptual shifts caused by varying an acoustic parameter. A biomechanical speech production model is proposed to explain the observed bifurcation and nonlinear behaviors [31]. The two-mass model indicates that bifurcation may be the result of desynchronization of left and right folds and the corresponding instabilities are similar to those observed in the voice of patients with unilateral paralysis.

A more complete dynamical analysis on speech signal is reported in [32], in which many invariant measures, such as singular value decomposition, Lyapunov exponents, and metric entropy, are computed. In [33], using exercised larynx experiments, bifurcation diagrams for vocal fold vibration are constructed to identify regions with distinct vibration patterns which are of physiological relevance. In addition to English, similar chaotic behaviors are also found in speech processes in other languages, such as Japanese vowels [34]. In fact, the nonlinear dynamics of speech has also triggered the application of chaos to other vocal systems. It is reported in [35] that the nonlinear phenomena of animal voices are common and are not limited to human beings or even mammals. In fact, at a different level of speech signals in the form of languages, nonlinear dynamics is found to provide a good language model [36].

As in many other chaotic analyses of real measured time series, different conclusions on the chaos properties of speech signals are reported [37]. Using sustained vowels from different speakers, it is observed that the speech signals used in the experiment are nonchaotic even though the signals possess low dimensionality. The nonchaotic conclusion is based on the use of a noise-robust Lyapunov exponent calculation algorithm. But in a more recent paper [38], contradictory observations are found in the sense that most of the speech signals used in [38] are shown to possess positive Lyapunov exponents. The authors do not conclude whether speech must be chaotic or not but point out that nonlinear dynamics does provide a useful tool for speech analysis and modeling, in particular, improved classification and prediction. In [39], the use of nonlinear dynamics in speech classification is further verified. Besides calculating the correlation dimension of measured speech signals, a theoretical model is developed to analyze the attractor dimension of speech signals. It is shown that nonlinear dynamics is effective in identifying pathological voices from patients with vocal tremor. In addition, the authors use high-speed imaging facilities to analyze the vocal vibration and conclude their observation on using chaos for speech identification [40].

Radar is another field that finds many applications of chaos in the literature. There are two main directions of chaos research in radar. One is modeling of natural radar signals such as radar backscattered signals from a sea surface by nonlinear dynamics. The other line of research is to design novel radar systems based on chaotic maps or circuits.

One of the earliest applications of chaos to radar signal processing is probably sea clutter modeling. It was first reported in [41] that radar backscattered signals from a sea surface, that is, sea clutter, has a fractional correlation dimension based on experimental sea clutter data. Similar to biomedical and speech signals, this observation initiates controversial investigations on modeling of sea clutter using chaos. [42] raised the concern whether the data size used in [41] is large enough to make such a conclusion. Around the same time, another independent study in [43] reported similar chaotic finding using different sea clutter data sets. More chaotic time series tests are applied to sea clutter, and chaotic behaviors such as positive Lyapunov exponents were reported [44, 45]. However, other research groups have a different conclusion on the same data base of sea clutter used in [44, 45]. They observed that colored random noise has a similar performance as sea clutter using the chaotic time series tests. [46] further concluded that sea clutter should be modeled as a linear autoregressive model with time-varying coefficients. But since a time-varying linear system can also be represented as a nonlinear system with constant coefficients, another research group [49, 50] continued this chaotic research and observed that sea clutter follows a power law sensitivities to initial conditions in a short time scale and the time scale has a strong impact on the chaotic behavior based on chaotic time series measurement. Whether sea clutter is truly chaotic or not and on what time scale remains controversial; many researchers do find that the nonlinear dynamical techniques provide useful tools for improved radar signal processing such as detection and classification.

As many radar systems such as noise radars require wideband signals, chaos becomes a suitable choice in such applications. There are two main approaches in this research. The first one is hardware design that can generate chaotic signals for radar systems. In [51], an optically injected semiconductor laser is used to generate chaotic signals with bandwidths larger than 10 GHz. This chaos-based radar is shown to have a very good range resolution compared to conventional systems. In [52], a Colpitts oscillator is applied to design chaos radar. Using the ambiguity functions, it is shown that this chaos radar has many sidelobes for detection but the cross-ambiguities functions indicate that it has a relatively good performance in electronic counter countermeasure. The second approach to chaos radar research is based more on designing modulation schemes or applying nonlinear dynamical processing techniques to radars. [6] applies chaos to random frequency modulated (FM) signals for wideband radar imaging and observes that the Bernoulli map has the best performance in their design. In [52], the chaos-based pulse amplitude modulation (CPAM) is proposed for ultra-wideband (UWB) radars. It shows very good clutter suppression capability and high resolution for through-the-wall imaging. In [53], by relating target parameters such as range and velocity with chaos generating parameters, it is shown that a high resolution noise radar can be designed using this principle. [54] used a spread spectrum radar based on chaotic spreading sequence for automotive applications. The use of chaos has shown potential in long range automotive radars. In [56], chaotic signals were used for complex targets and the transmitted signals are optimized for particular targets. A more recent work on chaos radar was reported in [57], in which the Lorenz-based chaotic signals are transformed to construct orthogonal waveforms for multiple-input multiple-output (MIMO) radars.

Among the various applications of chaotic signal processing, the most popular one should be communications. As chaotic signals have random behaviors, they can be used in different communication systems that require wideband signals, such as spread spectrum communications. After the seminal paper on the possibility in synchronizing chaotic systems [58], many researches have been dedicated to applying chaos synchronization concepts to communications. In [59], an analog circuit was developed for the Lorenz map to realize chaos synchronization in transmitting and receiving speech signals. Many different modulation schemes are then proposed to develop efficient chaos-based communications that may have potential for practical use. Chaos shift keying (CSK) is a popular approach which uses different chaotic waveforms to represent the digital or binary message for transmission [60]. While direct sequence spread spectrum (DSSS) uses pseudo noise (PN) as spreading sequences, some researchers suggest using chaotic signals to replace the PN sequences due to the good correlation properties of chaotic maps [61]. Another way to use a chaotic system for wideband communication is to embed the information message in some parameters of the chaotic system; this is called chaotic parameter modulation [62]. This modulation scheme is quite interesting from the signal processing viewpoint. While many schemes use correlators as their receivers as in conventional communication systems, the receiver design of chaotic parameter modulation can be treated as a problem of parameter estimation or tracking either the system parameters or the initial conditions, depending on how the information message is modulated [63]. In addition to using signal waveform or system parameters for modulation, the position of chaotic pulse has also been considered for chaotic pulse modulation [64]. With these different characteristics of a chaotic system for modulation, some of their extended versions are also developed including the differential chaos shift keying (DCSK) [65] and chaos-based DSSS with more realistic complex sequence implementation [66]. These extended schemes are found to have improved performance compared to their original versions and the performances are closer to conventional communication systems. For more details, the readers are referred to [67, 68].

Besides designing new chaos modulation schemes and analyzing their performances, another active area of chaos communication research is hardware implementation. The Lorenz map is first implemented electronically to validate the synchronization concept [59]. Another electronic circuit based on the double scroll oscillator is also considered for realizing chaos communication [69]. By controlling the symbolic dynamics, it is shown that communication based on chaotic circuit is feasible. Use of a laser is another approach for chaos communication hardware implementation. It is demonstrated in [70] that two chaotic semiconductor lasers can be synchronized. Using optical systems, chaotic signal is able to communicate at bandwidths of hundreds of megahertz, compared to the bandwidth of tens of kilohertz for chaotic electronic circuits [8]. An interesting experiment reported in [7] is that a chaotic laser can be used to transmit signals over 120 km of optical fiber in the metropolitan area network of Athens, Greece. It demonstrates the possibility of high-speed and long distance transmission based on chaos synchronization.

Signal processing is also found in various components of a communication system. In [71], chaos is applied to develop analog error correcting codes with a fast decoding technique as an alternative to conventional digital codes. [72] designs identically independent distributed p-ary random variables based on ergodic maps and applies these sequences to DSSS. In [73], a symbolic dynamic is considered for encoding information message. By analyzing the dynamics of coding, the authors conclude the existence of an optimal code from the viewpoint of channel capacity.

Other signal processing applications in chaotic communications in the literature include interference cancelation and channel equalization. By treating interference cancelation as an estimation problem in chaotic noise background, a dynamic-based estimator is applied to suppress narrow band interference in chaotic DSSS [74]. In [75], the co-channel interference minimization is achieved by minimization with respect to the autocorrelation function of some chaotic maps. Channel effect is another important issue in communications. In [76], chaos attractors are considered for multipath channel modeling and the performances of these chaos models are shown to be satisfactory in many cases. In [77, 78], chaos is applied in another way for channel effect. They treat equalization as an estimation problem with chaos as driving signals and exploit chaotic characteristics to develop effective schemes to mitigate channel effects.

More recent works on chaos communications include extending these chaos modulation techniques to current communication systems such as UWB communications [79, 80], wireless sensor networks [81], and MIMO systems [82]. These newer systems open up new problems that require chaotic signal processing techniques. In addition to the above mentioned chaos applications, other areas involving chaos that are of interest to signal processing researchers include security [9, 10], digital watermarking [73, 83], control [84, 85], computing [86], ranging [87], chemical sensing [88], and robotics [89, 90]. These applications may exploit chaos to achieve certain functions such as using the randomness property of chaos to design efficient search technique for mobile robots [90] and using the sensitivity property of chaos to enhance the sensing signals [88]. In summary, we observe that there are two main directions in chaotic signal processing. One is modeling that uses nonlinear dynamics analysis to describe natural signals such as speech, radar clutter, and EEG signals. The other is to apply generated chaotic signals and use their special characteristics such as correlation, wideband, or sensitivities to initial conditions in different systems such as communications, control, and radar to try to enhance system performance.

As we see from above, the application of chaos to engineering problems has drawn great attention and has been considered in many different disciplines. Although the concept is intriguing, chaos has not really been very successful in practical applications. In our opinion, one reason is that most of these chaos applications could not show any significant improvement or benefit when compared to the conventional approaches. For instance, while there is great interest in finding out whether a signal process such as speech and radar clutter is chaotic or not, there is a lack of good modeling or signal processing techniques that can take advantage of such a finding even if the signal process is indeed chaotic. Mostly, we will fit some nonlinear models such as neural networks to the signals, and this nonlinear model fitting does not really require that the signal be chaotic. In the other applications where generated chaotic signals are used and the signals being considered are truly chaotic, there is also lack of efficient processing techniques to fully exploit the chaotic properties. For instance, although chaos synchronization lays the foundation for chaos communication and many modulation schemes are proposed, an efficient receiver seems to be missing to show the advantage of chaos communication. In fact, many chaos communication systems use conventional concepts such as correlators which do not exploit any chaotic characteristics, and it is not surprising that no significant benefit is observed. Many researchers are aware of this problem, and many efforts have been dedicated to this signal processing research, as we will discuss in the following sections. However, it is noted that processing chaotic signals efficiently is a challenging problem, especially in noisy environments. For example, it is shown that estimating the parameter, more specifically the initial condition of a chaotic system from its measured signals corrupted by noise, suffers from a threshold effect [91]. It results in relatively large mean square errors (MSE) in the estimates when the signal-to-noise ratio (SNR) is below certain threshold and it explains the poor performance of some chaos applications such as chaotic parameter modulation for communication that requires estimating the parameters of a chaotic system.

In our opinion, to make these chaos applications practical, efficient chaotic signal processing techniques should be developed. These techniques should be able to exploit the underlying chaotic characteristics, especially in noisy environments. In the following sections, we will first give a formulation of chaotic signal processing, in particular, detection and estimation, as most signal processing problems can be categorized into these two functions. We will then present the detection and estimation techniques that have been developed in the literature. In the other chapters of this book, up-to-date research on detection and estimation techniques for chaotic signals is reported for various applications such as radar and communications.

1.2 ▪ Problem Formulation

As shown in the above section, a variety of problems involving chaotic signal processing arise in potential application contexts. Although the applications are different, they share similar signal processing formulation in their problems. In some cases, it is the chaotic signal that is of direct interest and is corrupted by random measurement noise. The related signal processing problems can be detection or estimation of chaotic signals in noise. In other cases, the chaotic signal acts as noise or interference in a detection and estimation problem, and signal processing is needed to extract the target signals in chaotic noise effectively.

Detection

We first consider detection of chaotic signals in noise. In this case, the signal of interest is chaotic, and the binary detection can be formulated as a binary hypotheses testing problem (H_1: target exists; H_0: no target). That is,

$$\begin{aligned} H_1: \quad & x(t) = s(t) + n(t), \\ H_0: \quad & x(t) = n(t), \end{aligned} \tag{1.1}$$

where $x(t)$ is the measured signal, $n(t)$ is the measurement noise and $s(t) = f(s(t-1),\ldots,s(t-M),\theta)$ is the chaotic signal generated by the nonlinear map f and θ represents the parameter vector of the map. It is noted that this formulation can also be applied to determine whether the measured signal is chaotic and two hypotheses can be referred as chaotic or random. For the more general classification problem, it can be extended as a multiple hypothesis testing problem:

$$H_i: \quad x(t) = s_i(t) + n_i(t), \quad i = 0,1,2,\ldots,N, \tag{1.2}$$

where $s_0(t) = 0$ represents the null hypothesis for background noise only. These N hypotheses are tested with $s_i(t) = f(s_i(t-1),\ldots,s_i(t-M),\theta)$ to determine which class the received signal should belong to.

If the measurement noise is Gaussian, the target signal can be treated as a random variable and the detection problem can be solved using conventional optimal decision theory. When the measurement is non-Gaussian, an optimal decision is more difficult to obtain. Different methods in the literature have been developed to address this problem. However, this statistical approach does not really take the chaotic nature of $s(t)$ into consideration. If the nonlinear function f is known, one can incorporate the dynamical equation into the likelihood function for the detection and classification problem. If f is unknown, a nonlinear prediction as an approximation of f can be used to exploit the nonlinear relationship among $s(t)$ and its previous values $s(t-1),\ldots,s(t-M)$ in the detection problem.

Besides working directly on the measured signals, detection and classification have also been considered at the feature level. Some features are first extracted from $x(t)$, that is, $\mu = g(x(t),\ldots x(t-M))$, where μ is usually some dynamical invariant such as correlation dimension and Lyapunov exponents. Since chaotic signals usually possess unique dynamical invariants, they can be used effectively to distinguish the chaotic and random hypotheses in (1.1). That is,

$$H_i: \quad \mu_i \text{ where } \mu_i \text{ is an invariant}, \quad i = 0,1,2,\ldots,N. \tag{1.3}$$

In fact, many chaos detection methods in the literature belong to this feature-based category, and many works reported above use this concept to detect whether chaos exists in measured signals such as EEG, speech, and radar clutter. The formulation in (1.3) is then used to classify different radar targets or different states of EEG signals.

The second situation is when the background noise $n(t)$ is chaotic, the binary test is the same as in (1.1). But this time $s(t)$ can be a random or deterministic signal and $n(t)$ follows a chaotic description $n(t) = f(n(t-1),\ldots,n(t-M),\theta)$. For classification, the problem is similar and the background noise $n_i(t)$ can be described by some nonlinear maps. In this detection scenario, the signal of interest is not chaotic, and it is the chaotic nature of $n(t)$ to be exploited for enhancing the detection of $s(t)$. For instance, we can use the invariant distribution of $n(t) = f(n(t-1),\ldots,n(t-M),\theta)$ in the likelihood function for hypothesis testing. Another more intuitive approach is to use the mapping $n(t)=f(n(t-1),\ldots,n(t-M),\theta)$ directly to describe the null hypothesis H_0. When the received signal $x(t)$ contains the target signal, it will not fit the mapping f, and hence larger errors will result. Hypothesis testing can then applied to the fitting error. That is,

$$\begin{aligned} H_1\colon\ & e(t) > \delta, \\ H_0\colon\ & e(t) < \delta, \end{aligned} \tag{1.4}$$

where $e(t) = (x(t) - f(x(t-1),\ldots,x(t-M),\theta)^2$ is the function fitting or prediction error and δ is the detection threshold. If the mapping f is unknown, as in many applications, then adaptive detection can be applied in the sense that background noise $n(t)$ is first extracted from the background so that a predictor f' can be developed to form an approximated model for $n(t)$. The predictor can be used in (1.4) to carry out dynamic-based detection.

Estimation

Similarly, we also classify chaotic estimation in two categories: estimating chaotic signals in noise and estimating deterministic or random signals in chaotic noise. When the signal of interest is chaotic, the estimation becomes a nonlinear state space estimation problem where the function in the state equation is nonlinear with parameters θ, and this nonlinear state equation does not require a noise term to drive it:

$$\begin{aligned} s(t) &= f(s(t-1), s(t-2), \ldots, s(t-M), \theta), \\ x(t) &= h(s(t)) + n(t), \end{aligned} \tag{1.5}$$

where $h(\cdot)$ is the measurement function. If θ is an unknown, then it becomes a problem of joint state and parameter estimation. Provided that the nonlinear mapping f is known, this estimation problem can be tackled by conventional state estimator such as the extended Kalman filter (EKF) or particle filter (PF). It is noted that the parameter θ may be the bifurcation parameters of the map or the initial condition $s(0)$. If f is known, estimating the initial value is equivalent to estimating the whole signal. In some situations, the functional form of f may not be available. One approach is to use a universal approximator $f'(\cdot;\eta)$ that can approximate any nonlinear function with properly selected parameters η. The estimation becomes a more complex joint state and parameter estimation problem given by

$$\begin{aligned} s(t) &= f'(s(t-1), s(t-2), \ldots, s(t-M), \eta), \\ x(t) &= h(s(t)) + n(t). \end{aligned} \tag{1.6}$$

The parameters η must be estimated in this case so that f' will be a close approximation of f.

When the background noise is chaotic, the formulation can be rewritten as

$$\begin{aligned} x(t) &= s(t,\alpha) + n(t), \\ n(t) &= f(n(t-1), n(t-2), \ldots, n(t-M), \theta). \end{aligned} \tag{1.7}$$

Although (1.7) looks similar to (1.5), the objectives of these two estimation problems are different. In (1.7), it is the parameter α of the target signal or the target signal $s(t,\alpha)$ to be estimated from the received signal $x(t)$. The state equation in (1.7) is not really needed in the estimation of $s(t,\alpha)$ unless that particular method incorporates the nonlinear map of the background noise in the estimation process. Since the chaotic process is now in the noise term $n(t)$, some chaotic characteristics of $n(t)$ such as entropy and finite dimensionality can be used as in the objective function of the estimation process instead of the conventional minimum variance or least squares as $n(t)$ is usually non-Gaussian.

In the following two sections, we survey some major chaotic signal processing techniques in the literature that have been developed to address the detection and estimation problems.

1.3 ▪ Detection Techniques

Detection is an important signal processing function found in many applications mentioned above. When the signals of interest are chaotic, as in speech and EEG applications, chaotic time series techniques [5, 18, 32] are applied to detect the presence of the signal and the dynamic invariants are measured for different categories of measured signals for classification and target recognition. Although these chaotic data analysis techniques can show the potential of chaos in these applications, they are not really designed for real-time signal processing and require very long data sequences for processing. A model-based approach was proposed in [92] to detect chaos from a short, noisy time series. It uses short time predictive power to discriminate chaos from random noise. Compared to some conventional tests for chaotic time series, it shows improvement in detecting chaos in noise using a shorter time series. Another feature besides predictive power that has been considered for detecting chaos is periodic unstable points. It was shown in [93] that low-dimensional chaos can be detected by extracting this feature and be distinguished from noisy limit cycles. However, the computation load of this approach is quite heavy. In [94], signal classification was carried out using another dynamic feature. It is shown that the phase space difference can distinguish different amplifiers if they are driven by the same or similar signals.

Conventional detection techniques in signal processing have also been considered for detecting chaos. In [95], statistical detection theory was applied to develop optimal detectors for chaos communications such as CSK and DCSK in additive white Gaussian noise. It is shown that a recursive implementation of the optimal detector takes the form of a Kalman filter. However, the likelihood derivation is in general quite challenging and is limited to some specific chaos maps only. In [96], a detection statistic was derived by searching for the phase space points in the attractor closest to the embedded noisy signal. This detector can be applied to general chaotic signals with improved performance compared to some conventional chaos communication receivers. Time-frequency analysis is another conventional signal processing technique being considered for chaos detection. By counting samples in the entire time-frequency domain, the chaotic states can be detected accurately even with a moderate level of noise [97]. In [98], machine learning was considered for detecting chaos. It uses a behavior tree to represent extracted patterns. Since chaotic time interval will never repeat in exactly the same way, it provides another way to detect chaos. In addition to these conventional signal processing techniques, there are methods that combine different techniques together for chaos detection. In [99], wavelet analysis was applied to decompose the EEG signals, and dynamical invariants such as correlation dimension and Lyapunov exponents are then extracted. A neural network is used to classify these extracted features, and this wavelet-chaos-neural network method is found to be quite accurate in classifying EEG signals. [100] presents another hybrid approach that combines chaos with centrum analysis. Using the observation that the phase transition is sensitive to the frequency difference of a periodic signal, it is shown that intermittent chaos is observed and the largest Lyapunov exponent has alternative symbols which can be used to address the shuttle valve early detection problem considered in [100].

Besides detecting chaotic signals in noise, chaos has also been considered in other detection applications. One interesting application is the use of chaos to enhance weak signal detection. In [101], a signal detection scheme based on the Duffing oscillator was proposed. It is observed that regular intermittent chaos happens if the frequency difference between the weak signal and the reference signal is small. An array of oscillators is used to cover the whole frequency domain to detect signals of unknown frequency. More detailed analysis of the probability distribution of the laminar length was given in [102] so that a more rigorous optimal detector can be applied. In other applications such as detection of signals in clutter where chaos is considered as noise [5, 44], a nonlinear predictor was used to model the background clutter, and then an adaptive model-based detector is formulated to detect signal embedded in the chaotic clutter background. By treating targets as the cause of changes in regular dynamics, the nonlinear predictive detection approach is capable of enhancing weak target detection in noise. Using a similar idea, a predictive model was constructed for network traffic and detection was performed based on the regular network traffic model to distinguish attacks from legitimate network traffic [103].

1.4 ▪ Estimation Techniques

Compared to detection, relatively more signal processing research has been dedicated to chaotic estimation. In fact, if the signal is chaotic, estimating the parameters of a chaotic signal can also be treated as a detection problem. Therefore, much work has been done on estimating parameters of a chaotic signal, either the bifurcating parameter or the initial conditions. One logical approach is to find the maximum likelihood (ML) estimates of the chaotic parameters as ML is an efficient estimator. In [104], a recursive ML estimator is developed for a one-dimensional chaotic map, namely, the tent map. This estimator is found to be asymptotically efficient at high SNR. The performance of the ML estimator is further analyzed in [105].

With a given map equation, estimating the signal is equivalent to initial condition estimation, and the Cramér–Rao lower bound (CRLB) of the initial condition estimate can be derived. It is shown that although the estimator is inconsistent, it is asymptotically unbiased and approaches the CRLB at high SNR. Assuming a one-dimensional unimodal map, a dynamic programming approach is developed to realize the ML estimates of the initial conditions in [106]. Because of the heavy computation load, a suboptimal method, called the halving method, was proposed in [106]. The initial condition of a unimodal map is determined from the itinerary of the map. It is a computationally efficient suboptimal approach and is further extended to general piecewise linear Markov maps in [107]. In addition to the special itinerary property, another approach that estimates the initial condition of a chaotic map, based on symbolic dynamics is proposed in [108]. Based on the relationship between the points in phase space and the admissible symbolic sequence, a fixed-lag smoothing algorithm is used to estimate the initial condition of a chaotic map, and the performance is found to be about the same as the halving method. This symbolic dynamic approach is further extended in [109]. It uses the symbolic dynamics in [108] to get an initial estimate and then refines the estimate by searching around its neighbor regions based on the sensitivity of a chaotic map. Recently, the initial condition estimation problem is investigated for spatial temporal dynamic systems [110]. Using the one-to-one relationship between the set of global orbits and the set of admissible sequences, the initial conditions of the coupled map lattices (CML) can be estimated from the symbolic sequences. The performance of initial condition estimation of CML is analyzed in [111]. It derives the CRLB and the condition on Lyapunov exponents for the CML initial condition estimation to be efficient.

Since many applications of chaos require real-time processing, sequential estimation is also considered for chaotic signals. The extended Kalman filter has been considered to estimate the states of the Lorenz system for chaos communication [112]. When the parameters of the chaotic system also need to be estimated, an augmented extended Kalman filter can be used to jointly estimate the states of the system and the system parameters [63, 113]. However, it is observed that the Kalman filter does not work effectively on chaotic systems as the Kalman gain does not converge for certain types of chaotic systems and exhibits an aperiodic phenomenon [114]. The mean square error of synchronization based on the extended Kalman filter is also derived in [115] to evaluate the performance caused by the aperiodic phenomenon. While other nonlinear filters such as the unscented Kalman filter are also found to suffer from this problem, exact filtering has been proposed for certain types of chaotic systems, such as logistic maps, to address this issue [116].

Considering estimating chaotic signals in noise as a state space estimation problem, system identification has also been a topic of interest to chaotic signal processing researchers. On one side, it is the chaotic system that we try to identify. In [117], the identifiability of a chaotic system is defined. Using chaotic synchronization, it is shown the possibilities of estimating the unknown parameters at the receiver from the measured signals. This identification problem is then extended to synchronizing chaotic systems, namely, the Lorenz map, with partial knowledge of the system parameters [118]. A detailed study on the stability of synchronization and parameter error dynamic is reported in [119]. It uses the persistent excitation property to construct an adaptive control algorithm to achieve synchronization and parameter estimation of chaotic systems.

On the other side, chaos can be used as a perturbation signal to drive a general system for identification. Since chaotic signals have impulses such as autocorrelation, the system is persistently excited when it is fed with such perturbation signals. A dynamic-based estimator is proposed in [120] to identify an autoregressive model. It is observed that a phase space volume of a chaotic signal is increased after passing system an autoregressive model, and hence the system parameters can be estimated by minimizing the phase space volume. Another approach to this identification problem is developed in [121], in which the predictability of a chaotic signal is exploited to estimate the system parameters by minimizing the prediction errors. An interesting theoretical result reported in [122] is that this semiblind or blind approach for system identification has the same theoretical performance as that of a nonblind approach. It shows the advantage of using chaos in signal processing. This system identification research is further extended to symbolic chaotic sequences in [123].

In addition to the relatively more standard estimators such as ML and Kalman filters, various approaches developed for chaotic estimation exploit different properties of chaos. In [124], symbolic dynamic is used to estimate the chaotic signals from its noisy measurement. It uses the similar partition concept in [108], but it reconstructs the chaotic signal of the one-dimensional piecewise linear Markov maps directly rather than estimating the initial condition. Another interesting result of using symbolic dynamics in chaotic estimation was reported in [125]. It shows how to estimate the parameter of the logistic map from the binary symbolic sequence when the initial value is unknown. However, the noise effect is not considered in their investigation. The unstable periodic orbit (UPO) is another interesting feature that has been considered for chaotic estimation. [126] was the first reported research on extracting UPO from time series measurements and then constructing a model from the extracted UPO for prediction. It shows interesting results compared to other prediction methods for chaotic signals, such as the local linear method. It was further shown in [127] that UPO has good performance in chaos-based DSSS as a receiver in terms of signal separation. Although UPO seems to be useful in dealing with strong measurement noise, it suffers from a very heavy computation load, as it requires search over a large number of UPO sequences.

One of the difficulties in processing chaotic signals is system nonlinearity. In [128], equivalent linear, random representations were derived for some chaotic systems such as the sawtooth map so that conventional signal processing techniques for linear systems can be applied. To address the threshold effect on chaotic estimation, an interesting approach was proposed in [129] for estimating a symmetric tent map. It constructs a set of Tent map sequences that allow consistent estimation and one can use these sequences for the engineering applications. [130] applied the ergodic theory to investigate some chaotic systems such as the sawtooth map and proposes a method to estimate the invariant density of piecewise affine maps. These theoretical results provide useful tools for applying statistical techniques to process chaotic signals. Recently, chaos was also considered for a newer signal processing topic, compressive sensing [131, 132]. Again, using the randomness property of chaos, chaos is used to construct the sensing matrix in compressive sensing as an alternative. More discussions on chaos for compressive sensing will be given in Chapter 7 of this book.

1.5 ▪ Summary

In this book, current researches on chaotic signal processing are reported in the following six chapters. They describe techniques for detection and estimation with applications ranging from radar and communications to computing. In Chapter 2, the problem of radar target recognition is considered. Instead of using chaos signal to design chaos radars, it investigates the use of nonlinear dynamics in processing radar returns to perform signal classification. Radar signals are processed in the phase space rather than the conventional time or frequency domain, and the phase space patterns are extracted for identification. Chapter 3 proposes a new approach for chaos communication. It uses a new class of chaotic oscillators that allow an exact analytic solution as a linear convolution of symbols with a fixed basis function. It is shown that this exactly solvable chaos system can theoretically achieve performance comparable to the conventional systems. Chapter 4 takes a different angle to process symbolic signals. It uses nonlinear dynamical systems to generate different logic functions to perform computing. This chaotic approach is shown to provide a reconfigurable computing device with improved fault tolerant. Chapter 5 addresses a conventional estimation problem, that is, system identification. It is shown that when the driving signal is chaotic, blind identification of a finite impulse response system is equivalent to nonblind identification, which makes using chaos for blind or semiblind identification an interesting alternative. In Chapter 6, the design of a high resolution LADAR system is described from a signal processing viewpoint. It is shown that the proposed system can generate clear images of objects such as aircrafts. A currently exciting topic in signal processing is compressive sensing. Chapter 7 provides an overview of this topic and the role of chaos in compressive sensing. These chapters report results from experts in chaotic signal processing. While some tend to develop improved signal processing techniques for chaos, some report up-to-date results on their specific applications such as radar, communications and computing to show the current status of chaos in engineering applications.

References

[1] A. Babloyantz and A. Destexhe, "Low-dimensional chaos in an instance of epilepsy," *Proc. Natl. Acad. Sci. USA*, 83, pp. 3513–3517, 1986.

[2] H. Adeli, S. Ghosh-Dastidar, and N. Dadmehr, "A spatial-temporal wavelet-chaos methodology for EEG-based diagnosis of Alzheimer's disease," *Neuroscience Letters*, 444, pp. 190–194, 2008.

[3] C. Ellegaard, K. Schaadt, and P. Bertelsen, "Acoustic chaos," *PhysicaScripta*, 90, pp. 223–230, 2001.

[4] X. Han, W. Li, and Z. Zeng, "Acoustic chaos and sonic infrared imaging," *Applied Physics Letters*, 81, pp. 3188–3190, 2002.

[5] H. Leung and T. Lo. "Chaotic radar signal processing over the sea," *IEEE Journal of Oceanic Engineering*, 18, pp. 287–295, 1993.

[6] B. C. Flores, E. A. Solis, and G. Thomas, "Assessment of chaos-based FM signals for range-Doppler imaging," *IEE Proc. Radar, Sonar, Navigation*, 150, pp. 313–322, 2003.

[7] A. Argyris et al., "Chaos-based communications at high bit rates using commercial fibre-optic links," *Nature*, 438, pp. 343–346, 2005.

[8] G. D. Van Wiggerren and R. Roy, "Communications with chaotic lasers," *Science*, 279, pp. 1198–1200, 1998.

[9] M. S. Baptista, "Cryptography with chaos," *Phys, Lett. A*, 240, pp. 50–54, 1998.

[10] L. Kocarev, K. S. Halle, K. Eckert, and L. O. Chua, "Experimental demonstration of secure communications via chaotic synchronization," *Internat. J. Bifur. Chaos Appl. Sci. Engrg.*, 2, pp. 709–713, 1992.

[11] J. P. Pijn, J. Van Neerven, A. Noest, and F. H. Lopes da Siva, "Chaos or noise in EEG signals: dependence on state and brain site," *Electroencephalography and Clinical Neurophysiology*, 79, pp. 371–381, 1991.

[12] W. J. Freeman, "Simulation of chaotic EEG patterns with a dynamic model of the olfactory system," *Biol. Cybernet.*, 56, pp. 139–150, 1987.

[13] J. Roschke and J. Aldenhoff, "The dimensionality of the human's electroencephalogram during sleep," *Biol. Cybernet.*, 64, pp. 307–313, 1991.

[14] J. Jeong, S. Y. Kim, and S. H. Han, "Nonlinear dynamical analysis of the EEG in Alzheimer's disease with optimal embedding dimension," *Electroencephalography and Clinical Neurophysiology*, 106, pp. 220–228, 1998.

[15] H. Adeli, S. Ghosh-Dastidar, and N. Dadmehr, "A wavelet-chaos methodology for analysis of EEGs and EEG subbands to detect seizure and epilepsy," *IEEE Trans. Biomedical Engineering*, 54, pp. 205–211, 2007.

[16] S. M. Pikkujamsa, T. H. Makikallio, et al.,"Cardiacinterbeat interval dynamics from childhood to senescence: comparison of conventional and new measures based on fractals and chaos theory," *Journal of the American Heart Association*, 100, pp. 393–399, 1999.

[17] J. Bhattacharya, "Detection of weak chaos in infant respiration," *IEEE Trans. Systems, Man, and Cybernetics, Part B*, 31, pp. 637–642, 2001.

[18] M. I. Owis, A. H. Abou-Zied, A. M. Youssed, and Y. M. Kadah, "Study of features based on nonlinear dynamical modeling in ECG arrhythmia detection and classification," *IEEE Trans. Biomedical Engineering*, 49, pp. 733–736, 2002.

[19] J. Holzfuss and W. Lauterborn, "Liapunov exponents from a time series of acoustic chaos," *Phys. Rev. A*, 39, pp. 2146–2152, 1989.

[20] I. Solodov, J. Wackerl, K. Pfleiderer, and G. Busse, "Nonlinear self-modulation and subharmonic acoustic spectroscopy for damage detection and location," *Applied Physics Letters*, 84, pp. 5386–5388, 2004.

[21] A. Mackay, "Application of chaos theory to ray tracking in ducts," *IEE Proc. Radar Sonar Navigation*, 146, pp. 298–304, 1999.

[22] A. L. Virovlyansky, "Statistical description of ray chaos in an underwater acoustic waveguide," *Acoustical Physics*, 51, pp. 90–100, 2005.

[23] T. Bodai, A. J. Fenwick, and M. Wiercigroch, "Ray chaos in underwater acoustics and its application," *Internat. J. Bifur. Chaos Appl. Sci. Engrg.*, 18, pp. 1579–1587, 2008.

[24] L. Fortuna, M. Frasca, and A. Rizzo, "Chaotic pulse position modulation to improve the efficiency of sonar sensors," *IEEE Trans. Instrumentation and Measurement*, 52, pp. 1809–1814, 2003.

[25] S. Azou, G. Burel, and C. Pistre, "A chaotic direct-sequence spread-spectrum system for underwater communication," *IEEE Oceans Conference*, 4, pp. 2409–2415, 2002.

[26] S. Azou, G. Purel, L. Le Deff, and C. Pistre, "Sea trial results of a chaotic direct sequence spread spectrum underwater communication system," *IEEE Oceans Conference*, 3, pp. 1539–1546, 2003.

[27] S. Kacimi and S. Laurens, "The correlation dimension: A robust chaotic feature for classifying acoustic emission signals generated in construction materials," *Journal of Applied Physics*, 106, pp. 1–8, 2009.

[28] W. Mende, H. Herzel, and K. Wermke, "Bifurcations and chaos in newborn infant cries," *Phys. Lett. A*, 145, pp. 418–424, 1990.

[29] B. Townshend, "Nonlinear prediction of speech," *IEEE Intl. Conf. Acoust., Speech, Signal Processing*, pp. 425–428, 1991.

[30] B. Tuller, P. Case, M. Ding, and J. A. S. Kelso, "The nonlinear dynamics of speech categorization," *Journal of Experimental Psychology: Human Perception and Performance*, 20, pp. 3–16, 1994.

[31] H. Herzel, D. Berry, I. Titze, and I. Steinecke, "Nonlinear dynamics of the voice: signal analysis and biomechnamical modeling," *Chaos*, 5, pp. 30–34, 1995.

[32] A. Kumar and S. K. Mullick, "Nonlinear dynamical analysis of speech," *Journal of Acoustical Society of America*, 100, pp. 615–629, 1996.

[33] D. A. Berry, H. Herzel, I. R. Titze, and B. H. Story, "Bifurcations in excised larynx experiments," *Journal of Voice*, 10, pp. 129–138, 1996.

[34] I. Tokuda, R. Tokunaga, and K. Aihara, "A simple geometrical structure underlying speech signals of the Japanese vowel /a/," *Internat. J. Bifur. Chaos Appl. Sci. Engrg.*, 6, pp. 149–160, 1996.

[35] W. T. Fitch, J. Neubauer, and H. Herzel, "Calls out of chaos: the adaptive significance of nonlinear phenomena in mammalian vocal production," *Animal Behaviour*, 63, pp. 407–418, 2002.

[36] W. G. Mitchener and M. A. Nowak, "Chaos and language," *Proc. R. Soc. London Ser. B*, 271, pp. 701–704, 2004.

[37] M. Banbrook, S. McLaughlin, and I. Mann, "Speech characterization and synthesis by nonlinear methods," *IEEE Trans. Speech and Audio Processing*, 7, pp. 1–17, 1999.

[38] I. Kokkinos and P. Maragos, "Nonlinear speech analysis using models for chaotic systems," *IEEE Trans. Speech and Audio Processing*, 13, pp. 1098–1109, 2005.

[39] Y, Zhang and J. J. Jiang, "Nonlinear dynamic mechanism of vocal tremor from voice analysis and model simulations," *Journal of Sound and Vibration*, 336, pp. 248–262, 2008.

[40] Y. Zhang, J. Shao, C. R. Krausert, S. Zhang, and J. J. Jiang, "High-speed image analysis reveals chaotic vibratory behaviors of pathological vocal fold," *Chaos Solitons Fractals*, 44, pp. 169–177, 2011.

[41] H. Leung and S. Haykin, "Is there a radar clutter attractor?" *Applied Physics Letters*, 56, pp. 593–595, 1990.

[42] D. Blacknell and C. J. Oliver, "Recognizing chaos in radar images," *J. Phys. D: Applied Physics*, 27, pp. 1608–1618, 1994.

[43] A. J. Palmer, R. A. Kropfli, and C. W. Fairall, "Signatures of deterministic chaos in radar sea clutter and ocean surface winds," *Chaos*, 5, pp. 613–616, 1995.

[44] S. Haykin and X. B. Li, "Detection of signals in chaos," *Proc. IEEE*, 83, pp. 95–122, 1995.

[45] S. Haykin and S. Puthusserypady, *Chaotic Dynamics of Sea Clutter*, John Wiley, 1999.

[46] C. P. Unsworth, M. R. Cowper, S. McLaughlin, and B. Mulgrew, "Re-examining the nature of radar sea clutter," *IEE Proc. Radar Sonar Navigation.*, 149, pp. 105–114, 2002.

[47] M. McDonald and A. Damini, "Limitations of nonlinear chaotic dynamics in predicting sea clutter returns," *IEE Proc. Radar Sonar Navigation*, 151, pp. 105–113, 2004.

[48] S. Haykin, R. Bakker, and B. W. Currie, "Uncovering nonlinear dynamics—the case study of sea clutter," *Proc. IEEE*, 90, pp. 860–881, 2002.

[49] J. Hu, J. B. Gao, and K. Yao, "Power law sensitivity to initial conditions in sea clutter," *IEEE 2005 International Radar Conference*, Arlington, VA, pp. 956–961, 2005.

[50] W. Tung et al., "On modeling sea clutter by noisy chaotic dynamics," *Proc. SPIE*, 6968, pp. G1–G10, 2008.

[51] F. Y. Lin and J. M. Liu, "Chaotic radar using nonlinear laser dynamics," *IEEE Journal of Quantum Electronics*, 40, pp. 815–820, 2004.

[52] Z. G. Shi et al., "Ambiguity functions of direct chaotic radar employing microwave chaotic Colpitts oscillator," *Progress in Electromagnetics Research*, 77, pp. 1–14, 2007.

[53] V. Venkatasubramanian and H. Leung, "A novel chaos-based high resolution imaging technique and its application to through-the-wall imaging," *IEEE Signal Processing Letters*, 12, pp. 528–531, 2005.

[54] Z. Liu, X. Zhu, W. Hu, and F. Jiang, "Principles of chaotic signal radar," *Internat. J. Bifur. Chaos Appl. Sci. Engrg.*, 17, pp. 1735–1739, 2007.

[55] E. Gambi, F. Choaraluce, and S. Spinsante, "Chaos-based radars for automotive applications: theoretical issues and numerical simulations," *IEEE Trans. Vehicular Technology*, 57, pp. 3858–3863, 2008.

[56] T. Carroll, "Optimizing chaos-based signals for complex radar targets," *Chaos*, 17, pp. 1–10, 2007.

[57] M. S. Willsey, K. M. Cuomo, and A. V. Oppenheim, "Quasi-orthogonal wideband radar waveforms based on chaotic systems," *IEEE Trans. Aerospace and Electronic Systems*, 47, pp. 1974–1984, 2012.

[58] L. M. Pecora and T. L. Carroll, "Synchronization in chaotic systems," *Phys. Rev. Lett.*, 64, pp. 821–825, 1990.

[59] K. M. Cuomo and A. V. Oppenheim, "Circuit implementation of synchronized chaos with applications to communications," *Phys. Rev. Lett.*, 71, pp. 65–68, 1993.

[60] H. Dedieu, M. P. Kennedy, and M. Hasler, "Chaos shift keying: modulation and demodulation of a chaotic carrier using self-synchronizing Chua's circuits," *IEEE Trans. Circuits Syst. II*, 41, pp. 634–642, 1993.

[61] G. Heidari-Bateni and C. D. McGillem, "A chaotic direct-sequence spread spectrum communication system," *IEEE Trans. Communications*, 42, pp. 1524–1527, 1994.

[62] T. Yang and L. O. Chua, "Secure communication via chaotic parameter modulation," *IEEE Trans. Circuits Syst. I*, 43, pp. 817–819, 1996.

[63] H. Leung and J. Lam, "Design of demodulator for the chaotic modulation communication system," *IEEE Trans. Circuits Syst. I*, 44, pp. 262–267, 1997.

[64] M. Sushchik et al., "Chaotic pulse position modulation: a robust method of communicating with chaos," *IEEE Communication Letters*, 4, pp. 128–130, 2000.

[65] G. Kolumban, M. P. Kennedy, Z. Jako, and G. Kis, "Chaotic communications with correlator receivers: theory and performance limits," *Proc. IEEE*, 90, pp. 711–732, 2002.

[66] G. Mazzini, G. Setti, and R. Rovatti, "Chaotic complex spread sequences for asynchronous DS-CDMA - Part I: system modeling and results," *IEEE Trans. Circuits Syst. I*, 44, pp. 937–947, 1997.

[67] F. C. M. Lau and C. K. Tse, *Chaos-Based Digital Communication Systems,* Springer, 2003.

[68] A. Abel and W. Schwarz, "Chaos communications – principles, schemes, and system analysis," *Proc. IEEE*, 90, pp. 691–710, 2002.

[69] S. Hayes, C. Grebogi, E. Ott and A. Mark, "Experimental control of chaos for communication," *Phys. Rev. Lett.*, 73, pp. 1781–1784, 1994.

[70] C. R. Mirasso, P. Colet, and P. Garcia-Fernandez, "Synchronization of chaotic semiconductor lasers: application to encoded communications," *IEEE Photonics Letters*, 8, pp. 299–301, 1996.

[71] B. Chen and G. W. Wornell, "Analog error-correcting codes based on chaotic dynamical systems," *IEEE Trans. Communications,* 46, pp. 881–890, 1998.

[72] T. Kohda, "Information sources using chaotic dynamics," *Proc. IEEE*, 90, pp. 641–661, 2002.

[73] E. Bollt and Y. C. Lai, "Dynamics of coding in communicating with chaos," *Phys. Rev. E*, 58, pp. 1724–1736, 1998.

[74] Y. Fu and H. Leung, "Narrowband interference cancellation in spread spectrum communication systems using chaos," *IEEE Trans. Circuits Syst. I*, 48, pp. 847–858, 2001.

[75] G. Mazzini, R. Rovatti, and G. Setti, "Interference minimisation by autocorrelation shaping in asynchronous DS-CDMA systems: chaos-based spreading in nearly optimal," *Electronics Letters*, 36, pp. 1054–1055, 1999.

[76] E. Costamagna, L. Favalli, and P. Gamba, "Multipath channel modeling with chaotic attractors," *Proc. IEEE*, 90, pp. 842–859, 2002.

[77] H. Leung, "System identification using chaos with application to equalization of a chaotic modulation system," *IEEE Trans. Circuits Syst. I*, 45, pp. 314–320, 1998.

[78] M. Ciftci and D. B. Williams, "Iterative equalization for chaotic communication systems," *IEEE International Conference on Acoustics, Speech, and Singnal Processing*, vol. 4, pp. 165–168, 2005.

[79] A. S. Dmitriev, M. Hasler, A. I. Panas, and K. V. Zakharchenko, "Basic principles of direct chaotic communications," *Nonlinear Phenom. Complex Syst.*, 4, pp. 1–4, 2002.

[80] C. C. Chong and S. K. Yong, "UWB direct chaotic communication technology for low-rate WPAN applications," *IEEE Trans. Vehicular Technology*, 57, pp. 1527–1536, 2008.

[81] G. Cimatti, R. Rovatti, and G. Setti, "Chaos-based spreading in DS-UWB sensor networks increases available bit rate,' *IEEE Trans. Circuits Syst. I*, 54, pp. 1327–1339, 2007.

[82] G. Kaddoum, M. Vu, and F. Gagnon, "On the performance of chaos shift keying in MIMO communications systems," *IEEE Wireless Communications and Networking Conference*, pp. 1432–1437, 2011.

[83] D. Zhao, G. Chen, and W. Liu, "A chaos-based robust wavelet-domain watermarking algorithm," *Chaos Solitons Fractals*, 22, pp. 47–54, 2004.

[84] E. Ott, C. Grebogi, and J. A. Yorke, "Controlling chaos," *Phys. Rev. Lett.*, 64, pp. 1196–1199, 1990.

[85] W. L. Ditto, S. N. Rauseo, and M. L. Spano, "Experimental control of chaos," *Phys. Rev. Lett.*, 65, pp. 3211–3215, 1990.

[86] S. Sinha and W. L. Ditto, "Dynamics based computation," *Phys. Rev. Lett.*, 81, pp. 2156–2159, 1998.

[87] K. Myneni, T. A. Barr, B. R Reed, S. D. Pethel, and N. J. Corron, "High-precision ranging using a chaotic laser pulse train," *Applied Physics Letters*, 78, pp. 1496–1498, 2001.

[88] F. Davide, M. Andersson, M. Holmberg, and I. Lundstrom, "Chaotic chemical sensing," *IEEE Sensors Journal*, 2, pp. 656–662, 2002.

[89] A. Goswami, B. Thuilot, and B. Espiau, "A study of the passive gait of a compass-like biped robot: symmetry and chaos," *The International Journal of Robotics Research*, 17, pp. 1282–1301, 1998.

[90] Y. Nakamura and A. Sekiguchi, "The chaotic mobile robot," *IEEE Trans. Robotics and Automation*, 17, pp. 898–904, 2001.

[91] I. Hen and N. Merhav, "On the threshold effect in the estimation of chaotic sequences," *IEEE Trans. Inform. Theory*, 50, pp. 2894–2904, 2004.

[92] M. Barahona and C. S. Poon, "Detection of nonlinear dynamics in short, noisy time series," *Nature*, 381, pp. 215–217, 1996.

[93] D. Pierson and F. Moss, "Detecting periodic unstable points in noisy chaotic and limit cycle attractors with application to biology," *Phys. Rev. Lett.*, 75, pp. 2124–2127, 1995.

[94] T. L. Carroll, "A nonlinear dynamics method for signal identification," *Chaos*, 17, 023109, 2007.

[95] M. Hasler and T. Schimming, "Optimal and suboptimal chaos receivers," *Proc. IEEE*, 90, pp. 733–746, 2002.

[96] G. K. Rohde, J. M. Nichols, and F. Bucholtz, "Chaotic signal detection and estimation based on attractor sets: Applications to secure communications," *Chaos*, 18, 013114, 2008.

[97] I. Djurovic and V. Rubezic, "Chaos detection in chaotic systems with large number of components in spectral domain," *Signal Processing*, 88, pp. 2357–2362, 2008.

[98] M. Nunez, R. Morales, and F. Triguero, "Automatic discovery of rules for predicting network management events," *IEEE Journal of Selected Areas in Communications*, 20, pp. 736–745, 2002.

[99] S. Ghosh-Dastidar, H. Adeli, and N. Dadmehr, "Mixed-band wavelet-chaos-neural network methodology for epilepsy and epileptic seizure detection," *IEEE Trans. Biomedical Engineering*, 54, pp. 1545–1551, 2007.

[100] Z. Zhao, F. Wang, M. Jia, and S. Wang, "Intermittent-chaos-and-ceptrum-analysis-based early fault detection on shuttle valve of hydraulic tube tester," *IEEE Trans. Industrial Electronics*, 56, pp. 2764–2770, 2009.

[101] G. Wang, D. Chen, J. Lin, and X. Chen, "The application of chaotic oscillators to weak signal detection," *IEEE Trans. Industrial Electronics*, 46, pp. 440–444, 1999.

[102] G. Wang and S. He, "A quantitative study on detection and estimation of weak signals by using chaotic Duffing oscillators," *IEEE Trans. Circuits Syst. I*, 50, pp. 945–953, 2003.

[103] A. Chonka, J. Singh, and W. Zhou, "Chaos theory based detection against network mimicking DDoS attacks," *IEEE Communications Letters*, 13, pp. 717–719, 2009.

[104] H. C. Papadopoulos and G. W. Wornell, "Maximum likelihood estimation of a class of chaotic signals," *IEEE Trans. Information Theory*, 41, pp. 312–317, 1995.

[105] S. Kay, "Asymptotic maximum likelihood estimator performance for chaotic signals in noise," *IEEE Trans. Signal Process.*, 43, pp. 1009–1012, 1995.

[106] S. Kay and V. Nageesha, "Methods for chaotic signal estimation," *IEEE Trans. Signal Process.*, 43, pp. 2013–2016, 1995.

[107] S. Wang, H. Leung, and P. C. Yip, "Estimating initial conditions of noisy chaotic signals generated by piece-wise linear Markov maps using itineraries," *IEEE Trans. Signal Process.*, 47, pp. 3289–3302, 1999.

[108] C. Ling, X. Wu, and S. Sun, "A general efficient method for chaotic signal estimation," *IEEE Trans. Signal Process.*, 47, pp. 1424–1428, 1999.

[109] L. Liu, J.Hu, Z. He, C. Han, H. Li, and J. Li, "Chaotic signal reconstruction with application to noise radar system," *EURASIP Journal on Advances in Signal Processing*, 2011:2, pp. 1–8, 2011.

[110] K. Wang, W. Pei, S. Wang, Y. M. Cheung, and Z. He, "Symbolic vector dynamics approach to initial condition and control parameters estimation of coupled map lattices," *IEEE Trans. Circuits and Syst. I*, 55, pp. 1116–1126, 2008.

[111] L. Lin, H. Shen, H. C. So, and C. Chang, "Convergence analysis for initial condition estimation in coupled map lattice systems," *IEEE Trans. Signal Process.*, 60, pp. 4426–4432, 2012.

[112] K. M. Cuomo, A. V. Oppenheim, and S. H. Strogatz, "Synchronization of Lorenz-based chaotic circuits with applications to communications," *IEEE Trans. Circuits Syst. I*, 40, pp. 626–639, 1993.

[113] D. J. Sobiski and J. S. Thorp, "PDMA-I: chaotic communication via the extended Kalman filter," *IEEE Trans. Circuits Syst. I*, 45, pp. 194–197, 1998.

[114] H. Leung, Z. Zhu, and Z. Ding, "An aperiodic phenomenon of the extended Kalman filter in filtering noisy chaotic signals," *IEEE Trans. Signal Process.*, 48, pp. 1807–1810, 2000.

[115] H. Leung and Z. Zhu, "Performance evaluation of EKF-based chaotic synchronization," *IEEE Trans. Circuits Syst. I*, 48, pp. 1118–1125, 2001.

[116] M. B. Luca, S. Azou, G. Burel, and A. Serbanescu, "On exact Kalman filtering of polynomial systems," *IEEE Trans. Circuits Syst. I*, 53, pp. 1329–1340, 2006.

[117] H. Dedieu and M. J. Ogorzalek, "Identifiability and identification of chaotic systems based on adaptive synchronization," *IEEE Trans. Circuits Syst. I*, 44, pp. 948–962, 1997.

[118] E. Solak, "Partial identification of Lorenz system and its application to key space reduction of chaotic cryptosystems," *IEEE Trans. Circuits Syst. II*, 51, pp. 557–560, 2004.

[119] A. Loria and A. Zavala-Rio, "Adaptive tracking control of chaotic systems with applications to synchronization," *IEEE Trans. Circuits Syst. I*, 54, pp. 2019–2029, 2007.

[120] H. Leung and X. Huang, "Parameter estimation in chaotic noise," *IEEE Trans. Signal Process.*, 44, pp. 2456–2463, 1996.

[121] Z. Zhu and H. Leung, "Identification of linear systems driven by chaotic signals using nonlinear prediction," *IEEE Trans. Circuits Syst. I*, 49, pp. 170–180, 2002.

[122] N. Xie and H. Leung, "Blind identification of autoregressive system using chaos," *IEEE Trans. Circuits Syst. I*, 52, pp. 1953–1964, 2005.

[123] N. Xie and H. Leung, "A nonlinear prediction approach for system identification using chaos symbolic dynamic," *IEEE International Conference on Systems, Man & Cybernetics*, Washington, D.C., pp. 1365–1370, 2003.

[124] J. Schweizer and T. Schimming, "Symbolic dynamics for processing chaotic signals I: noise reduction of chaotic sequences," *IEEE Trans. Circuits and Syst. I*, 48, pp. 1269–1282, 2001.

[125] X. Wu, H. Hu, and B. Zhang, "Parameter estimation only from the symbolic sequences generated by chaos system," *Chaos Solitons Fractals*, 22, pp. 359–366, 2004.

[126] K. Pawelzik and H. G. Schuster, "Unstable periodic orbits and prediction," *Phys. Rev. A*, 43, pp. 1808–1812, 1991.

[127] T. L. Carroll, "Spread-spectrum sequences from unstable periodic orbits," *IEEE Trans. Circuits Syst. I*, 47, pp. 443–447, 2000.

[128] D. F. Drake and D. B. Williams, "Linear, random representations of chaos," *IEEE Trans. Signal Process.*, 55, pp. 1379–1389, 2007.

[129] I. Rosenhouse and A. J. Weiss, "Consistent estimation of symmetric Tent chaotic sequences with coded itineraries," *IEEE Trans. Signal Process.*, 56, pp. 5580–5588, 2008.

[130] T. Addabbo, A. Fort, D. Papini, S. Rocchi, and V. Vignoli, "Invariant measures for tunable chaotic sources: robustness analysis and efficient estimation," *IEEE Trans. Circuits Syst. I*, 56, pp. 806–819, 2009.

[131] L. Yu, J. P. Barbot. G. Zheng, and H. Sun, "Compressive sensing with chaotic sequence," *IEEE Signal Processing Letters*, 17, pp. 731–734, 2010.

[132] D. V. Phong, N. Linh-Trung, T. D. Tan, H. V. Le, and M. N. Do, "Fast image acquisition in magnetic resonance imaging by chaotic compressed sensing," *IEEE International Symposium on Biomedical Imaging: From Nano to Macro*, Chicago, pp. 85–88, 2011.

Chapter 2

Target Recognition Using Nonlinear Dynamics

T. L. Carroll and F. J. Rachford

Abstract: Most analysis of radar reflections uses signal processing concepts based on random signals driving linear systems. We actually know far more about radar signals; radar signals are the output of dynamical systems, so we may use the assumptions of dynamics to extract more information from radar reflections. A radar transmitter is an example of a deterministic dynamical system; the behavior of the transmitter may be described by a finite number of rules, and transmitter behavior is reproducible: if started in the same state, the same behavior will result. In this chapter, we show that the concept of state space (or phase space) embedding may be used to analyze radar signals. Concepts developed for the measurement of the state space dimension are particularly useful for distinguishing radar signals reflected from different targets.

2.1 ▪ Introduction

It would seem that signals produced by chaotic systems would have many applications as radar signals. While there is a huge variety of possible chaotic signals, all chaotic signals have autocorrelation functions that decrease in time. This decreasing autocorrelation is a consequence of instability in the chaotic system, which causes small errors in measurement of a chaotic signal to increase exponentially with time, making deterministic chaotic signals predictable in theory but not in practice. Some chaotic signals can have broad bandwidths, giving them good range resolution for radar. Many groups have considered applying chaotic signals to radar problems [1, 2, 3, 4, 5, 6, 7, 8, 9].

With the advance of high speed digital to analog converters, however, the signal design problem has become much simplified. One may use numerical computation to optimize a signal for whatever application is desired and then output the signal through a high-speed arbitrary waveform generator. While it is possible that some of these designed signals may be chaotic, the presence of chaos isn't relevant for most applications, just the signal properties.

In this chapter, rather than looking at chaotic signals, we instead seek to analyze signals reflected from radar targets using recently developed methods from nonlinear dynamics. Because some of these new methods were developed with chaotic signals in mind, they look at signal properties in a very general way, without regard to whether the signals were generated by linear or nonlinear systems. The assumption that these new methods make is that the signals contain a finite number of degrees of freedom, i.e., they may be described by a system with a finite number of variables. An alternate way of stating this requirement is that the signals have a finite dimension.

We do consider one situation in which we use a chaotic system to generate signals for target identification, but in that case, the chaotic system isn't used for the chaos itself, but rather as a pseudorandom number generator that may be optimized. In the rest of this chapter, the signals used may or may not be chaotic; we are more interested in the signal analysis.

2.2 ▪ Radar

2.2.1 ▪ Point Targets

It is simplest to explain radar principles for a point target. In the simplest situation, we transmit a signal $v(t)$ which is reflected from a distant target. We receive the reflected signal $s(t)=v(t-\tau)+\eta(t)$, where $\eta(t)$ is a Gaussian noise signal and τ is a time delay. The value of τ tells us the distance to the target. Because $v(t)$ must travel to the target and back, for a target at a range of R, the amplitude of $s(t)$ has decreased by a factor of $1/R^4$ from the amplitude of $v(t)$, along with other factors having to do with target cross section, antenna gain, etc. [10]. Because the amplitude of the reflected signal is very small, noise is an important factor in detecting a radar signal. The basic problem of radar signal detection is this: given a transmitted signal $v(t)$ and a received signal $s(t)=v(t-\tau)+\eta(t)$, what is the probability of measuring the delay time τ? In terms of probability, we may state this as

$$P_v(\tau)=kP(\tau)P_\tau(v), \tag{2.1}$$

or the probability of measuring τ given received signal v is proportional to the prior probability that the delay actually is τ times the probability (or likelihood) that a delay value of τ results in the reflected signal s [11]. In order to maximize the probability of detection, we must maximize this last term. We may write $s(t)-v(t-\tau)=\eta(t)$ and use the fact that η is Gaussian to express the likelihood as

$$P_\tau(v)=\exp\left[-\frac{1}{N_0}\int(s(t)-v(t-\tau))^2dt\right], \tag{2.2}$$

where N_0 is the noise power. After some algebra, we find that the likelihood term becomes

$$R(\tau)=\int s(t)v(t-\tau)dt. \tag{2.3}$$

$R(\tau)$ is commonly known as the cross correlation.

For time varying signals, we may write $s(t) = \psi(t)\exp(2\pi i f_0 t)$ and $v(t) = \phi(t)\exp(2\pi i f_0 t)$, so $R(t)$ becomes

$$R(t) = \frac{\exp(-2\pi i f_0 \tau)}{N_0} \int \psi^*(t)\varphi(t-\tau)dt. \tag{2.4}$$

We may extract the envelope of $R(t)$ by summing its real and imaginary parts.

We would like to know the resolution with which we may determine a delay value τ_0. This amounts to measuring the width of $P_v(\tau)$ about τ_0. This width depends on the width of the envelope of $R(\tau) = P_\tau(v)$, Neglecting the noise term (2.4) gives

$$g(\tau) = \frac{1}{N_0} \int \varphi^*(t-\tau_0)\varphi(t-\tau)dt. \tag{2.5}$$

The maximum of $g(\tau)$ occurs at τ_0 and is equal to $2E/N_0$, where E is the signal power. Expanding $g(\tau)$ about τ_0,

$$g(\tau) = \frac{1}{N_0} \begin{bmatrix} \int |\varphi(t-\tau_0)|^2 dt + (\tau-\tau_0) \int \varphi^*(t-\tau_0)\frac{\partial}{\partial\tau}\varphi(t-\tau_0)dt + \\ \frac{1}{2}(\tau-\tau_0)^2 \int \varphi^*(t-\tau_0)\frac{\partial^2}{\partial\tau^2}\varphi(t-\tau_0)dt \end{bmatrix}. \tag{2.6}$$

To evaluate the integral we use the identity from Parseval's theorem,

$$d^n\varphi(t)/dt^n = 2\pi i f^n \Phi(f),$$

where $\Phi(f)$ is a coefficient in the Fourier expansion for $\varphi(t)$. Integrating over positive and negative frequencies, the first term in (2.6) is $2E$, where E is the total signal power, and the factor of two occurs because negative and positive frequencies are included. The second term in (2.6) is proportional to the average frequency of the signal $\varphi(t)$. In the complex representation of the signal, this average would be the carrier frequency, but in (2.6), we are using the low frequency, or baseband, signal, so the average frequency is zero.

The third term in (2.6) is proportional to the average of the square of the frequency. This average is used to define the bandwidth as β, where $2E\beta^2 = 2\pi i \int \Phi^*(f) f^2 \Phi(f) df$. Equation (2.6) then becomes

$$g(\tau) \approx \left(2\frac{E}{N_0}\right)\left[1 - \frac{1}{2}\beta^2(\tau-\tau_0)^2 + \cdots\right]. \tag{2.7}$$

From (2.1)

$$P_y(\tau) \propto \exp\left[-\frac{1}{2}(E/N_0)\beta^2(\tau-\tau_0)^2\right]. \tag{2.8}$$

Equation (2.8) looks like a Gaussian distribution with a standard deviation of

$$\delta\tau = 1\beta\sqrt{(E/N_0)},$$

so the width of the central peak in the cross correlation is inversely proportional to the signal bandwidth, or the accuracy with which the delay may be determined is proportional to the bandwidth (for small noise).

2.2.2 ▪ Ambiguity Function

If the target is moving, the determination of the distance to the target will be affected by the motion. Motion of the target along the line between the target and the radar receiver will expand or compress the reflected signal in time. For a signal containing only one frequency, the expansion or compression is equivalent to a frequency shift (the Doppler shift), but for more general signals, the expanded or compressed signal is [15]

$$s(t) = \sqrt{\gamma}v(\gamma t - \tau) + \eta(t), \\ \gamma = 1 + \frac{2v_t}{c}. \tag{2.9}$$

The probability of detecting the target as a function of delay and Doppler factor γ is therefore

$$|\chi(\tau,\gamma)|^2 = \left| \int_{-\infty}^{\infty} \sqrt{\gamma}v(t)v(\gamma t - \tau)\,dt \right|^2. \tag{2.10}$$

The ambiguity function is a probability distribution, so the volume under the surface given by the ambiguity function is a constant. The width of the peak of the ambiguity function along the τ axis is inversely proportional to the signal bandwidth, as shown in (2.8). For two signals having the same bandwidth and time length, the volume under the ambiguity function is the same [10, 11].

Because the volume under the ambiguity function is constant, improving the range resolution of a signal with a fixed time length and bandwidth, which results in a contraction of the central peak of the ambiguity diagram along the range axis, must result in the ambiguity diagram expanding along the Doppler axis, making the Doppler resolution worse. Improving Doppler resolution also results in worse range resolution. In practice, radar engineers may seek to "move ambiguity"; i.e., they may engineer signals that reduce the probability of false detection in certain parts of the range-Doppler space [12, 13], although this results in increasing the probability of false detection in other parts of the space. In this paper, adaptive chaotic maps will be used to "move ambiguity" on the ambiguity plot.

Figure 2.1 shows ambiguity functions for a point target for a chaos-based signal defined later ((2.17)–(2.18)), a linear frequency modulated (LFM) chirp with the same frequency and time span, and a sine wave frequency modulated with a uniformly distributed noise signal. The LFM chirp is a commonly used radar signal in which the signal frequency is swept continuously in a linear fashion between two values. The map that generated the chaotic signal has a largest Lyapunov exponent of 0.097 bits/iteration.

The length of each of the signals in Figure 2.1 was 8192 samples, and some additional processing was used with each signal before the ambiguity diagram was calculated as follows. The finite length of these signals imposes an additional modulation of the form of $\sin(x)/x$ on the cross correlation [10]. In order to minimize the finite length effects, each signal is first multiplied by a time window that has amplitude of one in the middle of the signal and small amplitude at the ends of the signal. The window in this case is the Blackman window [14], described by $w(t) = 0.42 - 0.5\cos(2\pi t/N) + 0.08\cos(4\pi t/N)$, where N is the number of samples in the signal.

The ambiguity diagram for the chaos-based signal has a large central peak, with smaller peaks about 15 dB below the main peak (15 dB is a factor of $10^{1.5}$, or approximately 30).

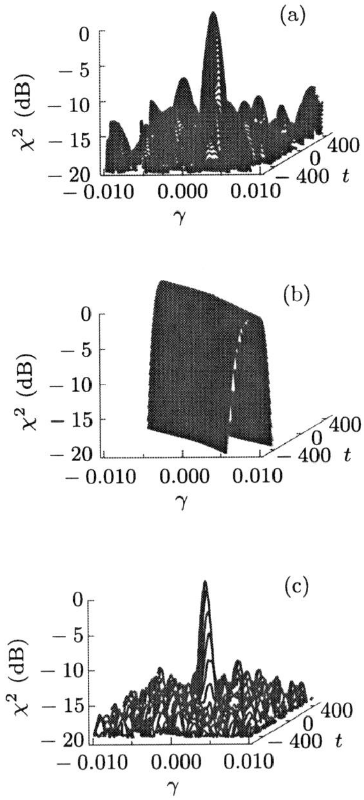

Figure 2.1. *Ambiguity functions. γ is the fractional Doppler shift (the amount by which the time axis is expanded), t is the delay in time steps, and χ is the value of the ambiguity function. The time step in the simulation was set as one, so for higher time steps, all factors will be multiplied by the appropriate scaling factor.* (a) *Ambiguity function for* 8192 *samples of a chaos-based signal described by* (2.17)–(2.18). *The map that generates the chaotic signal has a largest Lyapunov exponent of* 0.097 *bits/iteration.* (b) *Ambiguity function for a linear frequency modulated (LFM) chirp with the same frequency span and length as the chaos-based signal in* (a). (c) *Ambiguity function for a signal frequency modulated by uniformly distributed noise, with the same frequency span and time length as the signals in* (a) *and* (b). *Figure courtesy of US Navy.*

These side peaks will obscure a small target that is near the main target but whose amplitude is more than 15 dB below the amplitude of the main target. The side peaks are known as sidelobes, and low sidelobes are desirable. Figure 2.1(b) shows the ambiguity diagram for an LFM chirp, which is a commonly used radar signal because of its low sidelobes. The chirp does have a ridge in between the range and Doppler axes, which means that a Doppler shift is not distinguishable from a change in distance.

2.2.3 ▪ Complex Targets

If the radar target cannot be characterized as a point scatterer, much of the above discussion does not apply. The ambiguity function is not as useful, for example, because if the target presents a different aspect angle to the transmitter, the pattern of scatterers will change. This problem is made even more complicated by multiple scattering; the radar signal may reflect off several points on the target before returning to the transmitter. Resonances in the target cause by surface waves or internal cavities will also alter the target reflection.

2.2.4 ▪ Target Identification

A signal reflected from a radar target contains information about that target. How to extract this information has been the subject of much research [16]. One approach to radar target identification is to use a high range resolution signal to create an image of the target [16]. The image consists of a set of points for which the scattered signal has a high cross correlation with the transmitted signal. Image recognition techniques are then applied to recognize scattering patterns from different targets. This type of method is computationally intense, and because it relies on high resolution signals, the target signature varies greatly with the aspect angle.

In order to overcome this dependence on aspect angle, there are many target identification methods that depend on the fact that the target has resonances [17]. For simple shapes such as conducting spheres, for example, the incident radar signal induces creeping waves that travel along the surface of the sphere. When the creeping waves complete one circuit around the sphere, they are reradiated towards the transmitter, and for the right wavelengths, the reradiated waves interfere constructively or destructively with the secularly reflected signal. Waves may also reflect from discontinuities on the surface, or there may be resonant cavities within the target. Typically, to observe these resonances, a short broad band radar pulse is transmitted. This pulse excites many of the resonant modes in the target. After waiting for the specular reflection to pass, the transient ringing down of the target is recorded, and different methods may be used to extract the natural mode frequencies. Resonance based methods do not have the problem of aspect angle dependence, but usually the transient signal decreases in amplitude so that it is below the noise level in many situations, making frequency determination difficult.

Other target identification methods design optimum signals for particular targets [18]. If there are many possible targets, there are questions about how to efficiently design optimum signals; we present one method for addressing this problem later in this chapter.

2.3 ▪ Nonlinear Dynamics

The theory of dynamical systems studies systems that evolve deterministically over time [19, 20]. Because the systems are deterministic, there is some set of rules that govern the time evolution. Because they are governed by a system of evolution rules, the trajectories of dynamical systems form patterns. These trajectories can be described by a finite number of variables. Turbulent fluid flow or atmospheric dynamics are examples of dynamical systems with large numbers of variables, but for radar applications, we will consider only systems with no more than three or four variables.

A dynamical system based on ordinary differential equations may be described by

$$\frac{du}{dt} = f(u), \tag{2.11}$$

where u is a vector. A commonly studied dynamical system is the Lorenz system [21], described by

$$\begin{aligned}\frac{dx}{dt} &= 10(y-x),\\ \frac{dy}{dt} &= -xz+45x-y,\\ \frac{dz}{dt} &= xy-\frac{8}{3}z.\end{aligned} \tag{2.12}$$

The Lorenz system was originally developed to describe atmospheric dynamics. For the parameters used in (2.12), the Lorenz system is chaotic.

Plotting all three Lorenz variables in a phase space produces the trajectory shown in Figure 2.2. The Lorenz system is dissipative (it contains damping), so the motion of the Lorenz system occupies a finite region of phase space known as an attractor.

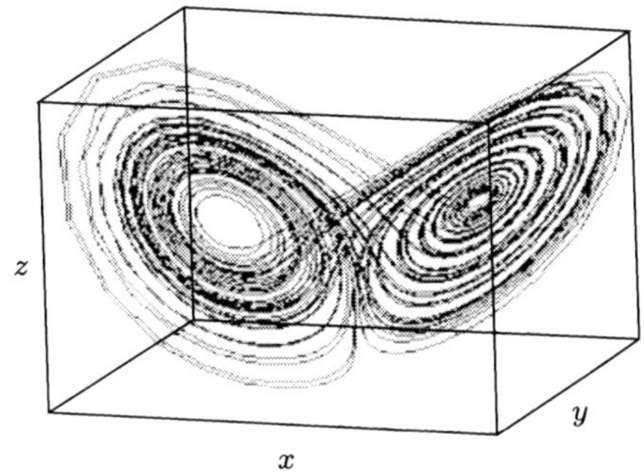

Figure 2.2. *Plot of the x, y, and z variables of the Lorenz system of* (2.12).

It is obvious from Figure 2.2 that the Lorenz attractor has a distinct structure. This structure is a consequence of the deterministic nature of the Lorenz system. Various properties of the Lorenz system may be calculated from the attractor of Figure 2.2, but in practice, for example in a radar system, we don't have access to all the variables of the dynamical system. Under certain conditions it has been proven that we may use the one signal only from a dynamical system to construct an object that has the same topological properties as the original attractor [22]. If we have one signal from a dynamical system, $x(t)$ for example, we use the method of delays to construct a set of vectors from $x(t)$ and delayed versions of $x(t)$:

$$x(t)=\left[x(t),x(t+\tau_1),x(t+\tau_2)\ldots x\left(t+\tau_{d-1}\right)\right]. \tag{2.13}$$

The number of delays is equal to the dimension d of the delay vector. If there are N variables in the original dynamical system, then the dimension of the delay vector necessary to represent the dynamics is $d \leq 2N+1$. The attractor reconstructed in this way is an embedding [23] of the original attractor. The actual number of dimensions required for the embedding and the necessary delay values may be determined by known methods [24, 25]. These delay embedding methods haven't been proven for signals that have been modulated in frequency or amplitude, but they appear to give useful results for the radar signals studied here.

The properties of the attractor that we will be interested in for this chapter are the Lyapunov exponents and the dimension of the attractors. Starting from (2.11), we may calculate the matrix of T of partial derivatives of the function f, where

$$T^t_{ij} = \partial f[u_i(t)]/\partial f[u_i(t)]\partial u_j(t).\partial u_j(t).$$

The Lyapunov exponents λ are then defined as [26]

$$\lim_{t\to\infty} \frac{1}{t} \log T^t_i u(t) = \lambda^{(i)}. \tag{2.14}$$

The physical meaning of the Lyapunov exponents is that if we make a small perturbation $\delta u(0)$ in the initial conditions of a dynamical system, then $\delta u_i(t) = \delta u(0)e^{\lambda^{(i)}t}$. A continuous chaotic system such as the Lorenz system has a minimum of three Lyapunov exponents: one positive, one zero, and one negative.

2.4 ▪ Adaptive Maps for Target Identification

Although we are mostly interested in signal analysis and not signal generation, we do mention here a situation in which a chaotic system is used to generate signals. Because chaotic system can produce a huge variety of broad band signals, we use a chaotic map as an optimizable pseudorandom signal generator. The chaotic map requires only a few parameters to generate a large variety of signals, so it may be used to increase the speed of an optimization problem [27, 28].

2.4.1 ▪ Chaos versus Noise

A well-known result from radar and sonar analysis states that an LFM chirp signal will have a better signal to interference plus noise ratio (SINR) in the presence of interference than a noiselike signal [29]. This analysis is based on the properties of the ambiguity diagram. One can consider a group of scatterers distributed with a uniform density over a fixed area (the interference). The distribution can be along both the range and Doppler axes. A schematic representation of this interference is shown as the shaded area in Figure 2.3. The amount of interference produced by the scatterers will be proportional to the volume that is under the intersection of the ambiguity surface and the scattering area. The ambiguity plot for the LFM chirp (Figure 2.1(b)) is large only over a small area, so the volume that is also within the scattering area will in general be small, as can be seen in Figure 2.3(a); the LFM chirp ambiguity is large only within the diagonal ellipse. The noise ambiguity plot (Figure 2.1(c)) has a large central peak, surrounded by an area of relatively uniform and nonnegligible amplitude, shown as a large circle in Figure 2.3(b). From the analysis in Stewart and Westerfield [29], the volume under the intersection of the ambiguity surface and the interference area is larger for the noise signal than for the LFM chirp, so the chirp has a higher SINR than the noise signal.

The above analysis does not apply to chaos because chaos is not noise. While a time series signal generated by a chaotic system is unpredictable, there are recurrent patterns within chaotic signals, even broad band signals. A signal from a particular chaotic system has a grammar, where some patterns are allowed and some patterns never occur [30, 31]. Patterns in an actual chaotic signal may be seen through the use of recurrence plots [32].

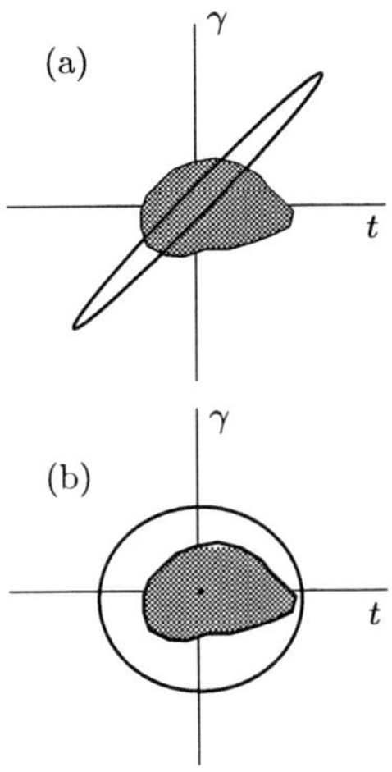

Figure 2.3. *Schematic representation of the ambiguity plots for an LFM chirp* (a) *or noise* (b) *in the presence of interference (shaded area).* γ *is the Doppler shift axis, and* t *is the time axis. Figure courtesy of US Navy.*

A random signal will contain every possible pattern with equal probability, while some patterns may not be seen in a chaotic signal (or the probabilities may not be equal), so the ambiguity surface for a chaos-based signal may not be uniform, as it is for noise.

As an example, the ambiguity surface for the chaos-based signal in Figure 2.1(a) is much less uniform than the ambiguity surface for the noise-based signal in Figure 2.1(c). These nonuniform areas can be moved around by choosing different chaos-based signals, so in principle, for a particular set of scatterers, it could be possible for the chaos-based signal to have a higher SINR than a chirp signal. One objective of this paper is to look for such chaos-based signals.

In this section, in order to identify targets, we cross correlate the signal reflected from a target with a reference signal which has not been transmitted. Because the target is extended in space, we cross correlate the reflected signal with the reference signal and then average the cross correlation across the target width. For two targets, A and B, the signal reflected from target A is $s(A)$, while the reflection from B is $s(B)$. We look for a reference signal $S_{\max}(A)$ that maximizes the average correlation with $s(A)$ while simultaneously minimizing the average correlation with $s(B)$. We also look for the signal $S_{\min}(A)$ that minimizes the average correlation with $s(A)$ and maximizes the correlation with $s(B)$. We repeat this process to get $S_{\max}(B)$ and $S_{\min}(B)$.

Denoting the correlation between x and y as $R(x,y)$, if we have an unknown reflection $s(u)$, we can calculate the contrast ratio $C(A,s(u))$ with the reference signals for target A:

$$C(A,s(u)) = 20\log_{10}\left(\frac{R(S_{\max}(A),s(u))}{R(S_{\min}(A),s(u))}\right). \tag{2.15}$$

We may calculate a corresponding contrast ratio $C(B,s(u))$ for target B. We then decide that the signal $s(u)$ was reflected from target A if $C(A,s(u)) > C(B,s(u))$.

In general we have many targets to distinguish, so we maximize

$$G = \frac{R(S_{\max}(A), s(A))}{R(S_{\min}(A), s(A))} + \sum_{i=1}^{N} \frac{R(S_{\max}(A), s(A))}{R(S_{\max}(A), s(i))} + \sum_{i=1}^{N} \frac{R(S_{\min}(A), s(i))}{R(S_{\min}(A), s(A))}, \tag{2.16}$$

where there is a total of N targets.

It is possible to do this optimization simply by varying the spectral components of the signal $S_{\max}(A)$. If we have a pulse that is T_l seconds long and has a bandwidth of W Hz, then we need to sample at a rate of $T_s = 0.5/W$. We then have a total of $N_t = T_l/T_s$ digitized points, so we have $N_o = 2N_s$ spectral components to optimize (we have real and imaginary parts). For example, if we have a 1 μs pulse with a 50 MHz bandwidth, we have 200 variables to optimize.

As an alternative to this optimization, we introduce a 6-parameter chaotic map:

$$\begin{aligned} x_{j+1}(1) &= \left[\sum_{i=1}^{6} p(i)x_j(i)\right] \bmod 1, \\ x_{j+1}(i) &= x_j(i+1) \qquad (i = 1,2,\ldots 5), \\ x_{j+1}(6) &= x_j(1), \\ \xi(j) &= x_j(1) + 0.5. \end{aligned} \tag{2.17}$$

Setting the number of parameters at 6 is arbitrary. The map parameters $p(i)$ are adjustable.

The map output signal $\xi(j)$ is then modulated onto a carrier. The carrier signal is a set of concatenated sinusoids, with the frequency of the j'' sinusoid set by $\xi(j)$:

$$v(t) = \sin(2\pi t/[20\xi(j)]) \qquad [t = 0,1,2,\ldots,t < 20\xi(j)]. \tag{2.18}$$

Different signals could be produced by varying the map parameters $p(i)$. By varying the map parameters, we could optimize G in (2.16). Instead of 200 parameters, as in the example above, there were only 6 parameters to optimize. This optimization works because the chaotic map can generate a huge variety of signals; we are essentially using the map as an optimizable pseudorandom code generator. This map could generate periodic as well as chaotic signals, so in the optimization, the parameter set was constrained to produce only chaotic signals by enforcing

We applied this optimization procedure to reflections from simulated missiles [33]. The simulated missiles are shown in Figure 2.4. Target 14 was 4.9 m long, while target 16 was 4.8 m long, and both targets contained features such as engine ducts and radomes. Both models had a 2 m wingspan. We used a finite difference time domain (FDTD) code to launch linear chirps at the models at various angles in one degree steps between zero and 180°. The chirps had a center frequency of 250 MHz and a bandwidth of 50 MHz. The small bandwidth was chosen to minimize the variation of target cross section with aspect angle. The same chirp was used for each simulation. The monostatic return waveform was calculated in the far field for each angle, and the FDTD simulation time was limited to minimize the buildup of rounding errors. The target size and waveform length were limited by simulation constraints.

The chirp was constructed by concatenating 33 individual sinusoids with individual frequencies from 225 to 275 MHz. In order to minimize aspect angle variations, the reference waveforms were optimized for signal returns obtained over 5° angular windows.

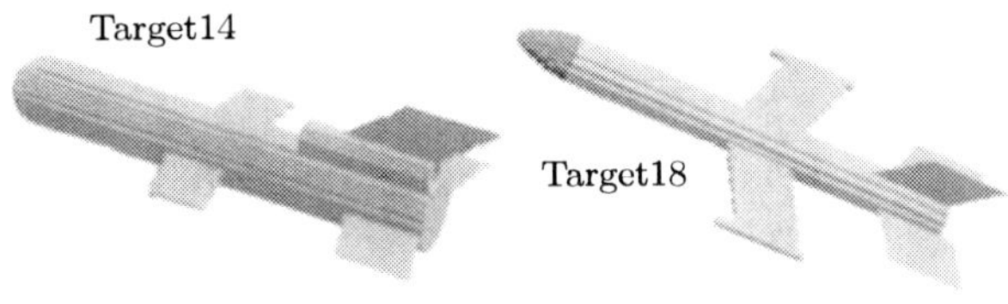

Figure 2.4. *Simulated missiles used to test the target identification algorithm. Figure courtesy of US Navy.*

A downhill simplex optimization algorithm was used [34]. The optimization problem was nonlinear, so the optimization routine was started from a number of different initial conditions, and resulted in a number of local minima. The waveform corresponding to the lowest local minimum was used, although in most cases there were several minima that were approximately equivalent.

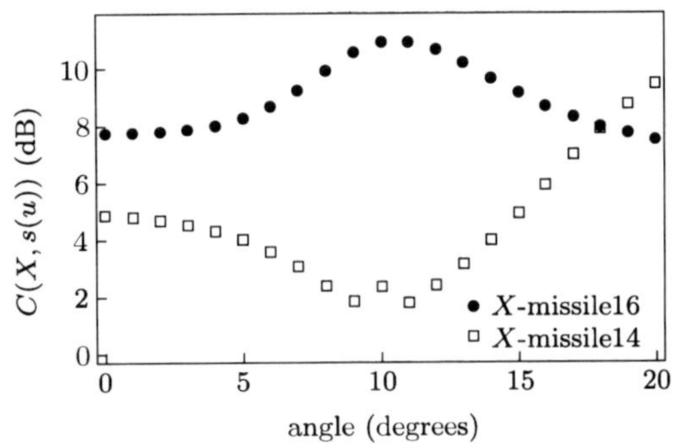

Figure 2.5. *Contrast ratios* (2.15) *for a signal* $s(u)$ *that has been scattered from missile* 16. *The contrast ratios were optimized for an angle of* 10°. *For most of the angles shown, the contrast ratio is larger for missile* 16, *indicating a correct target identification. The contrast ratio curves vary smoothly, indicating that the target identification varies only slowly with aspect angle.*

Figure 2.5 shows that the contrast ratios of (2.15) correctly identify which missile an unknown signal was scattered from. The contrast ratios were optimized for an aspect angle of 10°, but the contrast ratios still give the correct identification for a broad range of angles on either side of the optimum, indicating that this target identification method is not overly sensitive to aspect angle.

2.5 ▪ Signal Processing Methods

Most of this chapter is not about the signals themselves but about applying signal processing methods based on nonlinear dynamics to radar problems. Rather than looking at signals in the time or frequency domains, we begin by embedding signals in phase space, as described in (2.13). We then study one small part of this phase space at a time by choosing some index point on the embedded attractor and searching for other points on the attractor that are nearby in phase space. There are a number of search algorithms available; for our work, we use the slice search algorithm [35] for its ease of implementation.

A group of nearby neighbors in phase space represents a set of similar initial conditions for the underlying dynamical system. Starting a dynamical system multiple times from the same initial conditions will result in identical motion, so starting the dynamical system from multiple initial conditions that are close in phase space will initially result in motion that is similar. We may use this similarity of motion to characterize the average dynamics over a small region of phase space. If the dynamical system is chaotic, trajectories resulting from different initial conditions will diverge because of the positive Lyapunov exponent (defined in (2.14)).

2.5.1 ▪ Improving Detection Accuracy

We may use the positive Lyapunov exponents of a chaotic signal to improve the accuracy with which we locate a target. As shown in (2.8), the width along the range axis of the central peak of the ambiguity function for a signal (2.10) is determined by the bandwidth of the signal. While this accuracy limit is true only for detection using cross correlation, the cross correlation gives the highest probability for signal detection in Gaussian noise, so we don't want to throw out the cross correlation. Instead, we choose a different signal to correlate.

We denote the signal that we transmit as $x(i)$ and the received signal as $y(i)$. We keep a copy of $x(t)$ at the transmitter as a reference. We begin by embedding both $x(t)$ and $y(t)$ in a phase space with the proper dimension and delay values [25] to produce the vectors $x(i)$ and $y(i)$.

We then choose a reference point i_x on the embedded signal $x(i)$. Usually we begin by choosing the first point in the signal, but we may also choose a random index point. We then search $y(i)$ for the point $y(i_y)$ that is closest in phase space to $x(i_x)$. The two indices, i_x and i_y are used to create two new time series: $x_y(i) = x(i_y)$ and $y_x(i) = y(i_x)$. This procedure amounts to exchanging indices between the two time series. We then choose a new index point i_x and repeat this process, ending up with two new time series, $x_y(i)$ and $y_x(i)$.

If the signals $x(i)$ and $y(i)$ are aligned in time, then $x(i_y)$ will be very close to $y(i_x)$; if the two signals are not aligned, then $x(i_y)$ will not be close to $y(i_x)$. The distance between these points will depend on the time delay between $x(i)$ and $y(i)$, and on how fast the two signals diverge in phase space. The rate of divergence depends on the largest Lyapunov exponent for the chaotic system (2.14).

We then measure the cross correlation $R(x_y, y_x)$ between $x_y(i)$ and $y_x(i)$. Depending on the value of the largest Lyapunov exponent for the system that generated x and y, the accuracy with which $R(x_y, y_x)$ determines the delay between x and y may be greater than the accuracy of $R(x, y)$. We have plotted the ambiguity functions for two different signals, using either $R(x, y)$ or $R(x_y, y_x)$ in Figures 2.6 and 2.7.

Figure 2.6 shows ambiguity functions for a linear frequency modulated chirp signal. Using a time step of 1, the chirp was 10,000 points long, with frequencies from 0.08 to 0.12. Figure 2.6(a) shows the ambiguity function calculated by comparing x and y, while

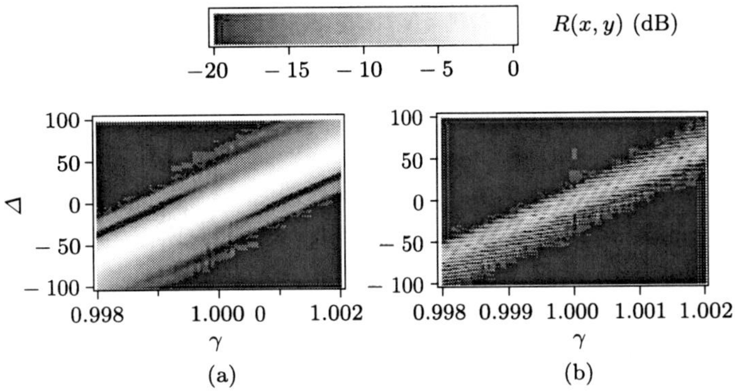

Figure 2.6. *Ambiguity functions for a linear chirp signal, calculated either the normal way* (a) *or using the signals* x_y *and* y_x *extracted from phase space* (b). *The axis* γ *is the Doppler factor* (2.9), *while* Δ *is the time shift in number of points*

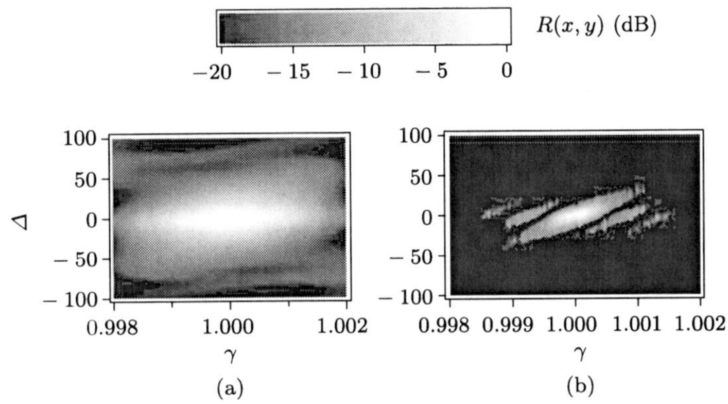

Figure 2.7. *Ambiguity functions for a chaotically frequency modulated signal, calculated either the normal way* (a) *or using the signals* x_y *and* y_x *extracted from phase space* (b). *The axis* γ *is the Doppler factor* (2.9), *while* Δ *is the time shift in number of points*

Figure 2.6(b) is calculated using x_y and y_x. The ambiguity functions were calculated using a Welch window. The central peak in range of the ambiguity function is narrower when x_y and y_x were used, and the range sidelobes are lower. A chirp is not actually a chaotic signal, but nearby trajectories diverge for a chirp signal embedded in a phase space, so the chirp has some properties in common with a chaotic signal.

In Figure 2.7, the ambiguity function was calculated for a chaotically frequency modulated signal, generated by the process of (2.18). The chaotic map in this case was $x_{n+1} = 2.1x_n \mod 1$, and the signal $\xi(j) = 0.2x_n + 0.9$. The resulting signal was 10,000 points long.

In Figure 2.7, $R(x_y, y_x)$ is narrower than $R(x, y)$ in both range and Doppler axes. Using the positive Lyapunov exponent of the chaotic system allows us to get greater accuracy in measuring the location of a target.

2.5.2 ▪ Target Detection from Patterns in Phase Space

Although a positive Lyapunov exponent would seem to be the characteristic feature of chaos, the nonlinearity of a chaotic system is also important in creating the unique features of a chaotic signal. Without some form of nonlinearity, the motion of a chaotic signal would continue to expand without bound along the direction of the positive Lyapunov exponent. Some type of nonlinearity is required in a chaotic system to keep the motion bounded.

The interaction between expansion and folding produces a fine structure in a chaotic attractor. This structure is similar on all length scales, so it is fractal. When a chaotic signal is scattered from a radar target or passed through a filter, this fine phase space structure may be used to identify the target. We may use finite impulse response (FIR) filters as a simple model for a radar target; an FIR filter consists of a weighted sum of delayed versions of an input signal, which is similar to what happens to a signal scattered from a radar target.

2.5.3 ▪ Phase Space Structures

We may use the Hénon map to show a simple example of the complex fine structure that is created in phase space. The Hénon map is given by [36]

$$\begin{aligned} x(n+1) &= y(n)+1-\alpha x(n)^2, \\ y(n) &= \beta x(n). \end{aligned} \tag{2.19}$$

For this example, we used the x signal from the Hénon map of (2.19), with $\alpha = 1.29$ and $\beta = 0.3$.

From a 25,000 point time series of x we pick an index point $x(i_0), x(i_0+1)$ and find the 40 closest neighbors. We plot these neighbors in Figure 2.8(a). There is a noticeable structure to this plot, which is expected, because the Hénon system is two dimensional.

We then create a simple filtered signal: $z_1(t) = x(t) + x(t+10)$. We pick an index point $z_1(i_0), z_1(i_0+1)$ and find its 40 closest phase space neighbors. These points are plotted in Figure 2.8(b). The signal z_1 was produced from x by an FIR filter, so it is still two dimensional, but the dimensionality is not obvious from the complex phase space plot. Filtering of x has produced a fine structure in phase space, so it is possible that the points that we have located as phase space neighbors on z_1 are not really the closest neighbors.

Using a different filter produces a very different structure. We created a second filtered signal, $z_2(t) = x(t) + x(t+11)$. In Figure 2.9, we plot the same 40 phase space neighbors from $z1$ as in Figure 2.8(b), but we also plot points from z_2 that have the same indices as the phase space neighbors as z_1. The points from z_2 are obviously not neighbors, and they are spread out over a much larger region in phase space than the points from z_1. This difference suggests a possible method to distinguish different filters; using the same set of indices, points that are phase space neighbors on one filter may not be neighbors on a different filter.

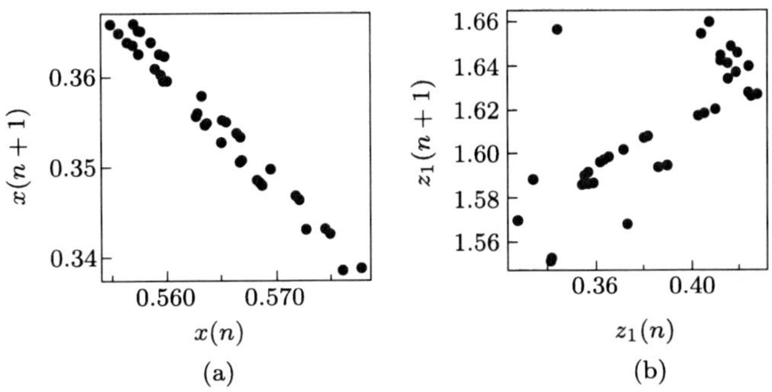

Figure 2.8. (a) *Nearest phase space neighbors for an arbitrary phase space point from the x signal of the Hénon map of* (2.19). (b) *Nearest phase space neighbors from an FIR filtered signal* z_1*, which is a filtered version of the x signal.*

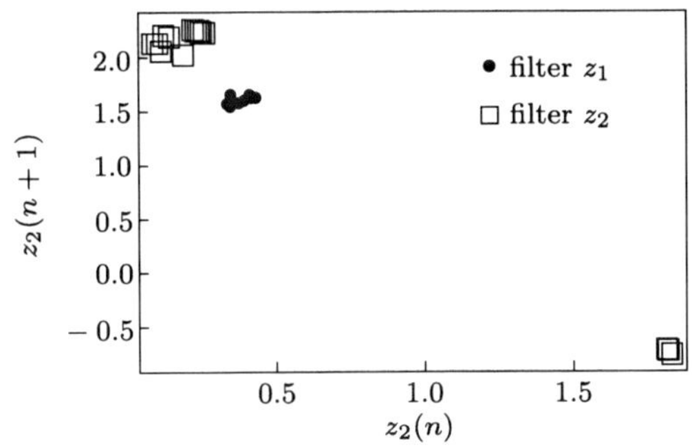

Figure 2.9. *Nearest phase space neighbors for an index point from the filtered signal* z_1 *and points with the same indices from a different filtered signal* z_2.

2.5.4 ▪ Dimension

The complex structure in phase space created by filtering a chaotic signal looks like an increase in the signal dimension, although it may or may not be a real dimension change. It has been known for some time that filtering a chaotic signal may alter its dimension [37]. Dimension in this case refers to the dimension of the chaotic signal measured in a phase space (or state space) [24]. The dimension of the signal will be less than or equal to the number of variables in the dynamical system that generates the signal. How the dimension increase comes about may be demonstrated by calculating the Kaplan–Yorke dimension [38]:

$$D_L = j + \sum_{k=1}^{j} \frac{\lambda_k}{\left|\lambda_{j+1}\right|}, \tag{2.20}$$

where the λ's are the Lyapunov exponents [39] for the chaotic system and j is the largest index for which the sum over k is nonnegative. The Lyapunov exponents describe the initial rate of divergence of nearby trajectories in a chaotic system. If a chaotic signal is filtered by a recursive (IIR) filter, the negative Lyapunov exponent from the filter is added to the Lyapunov exponent spectrum of the chaotic signal, possibly increasing the Kaplan–Yorke dimension.

This dimension increase can be a signature for a target that acts as an IIR filter. Theoretically, this dimension increase happens only for IIR filters, but in practice, an FIR filter, which is not recursive, may also appear to cause a dimension increase [40, 41]. As the order n of an FIR filter approaches ∞, the filter itself approaches an IIR filter [42]. Filtering of a chaotic signal by an FIR filter will introduce fine structure into the phase space of the chaotic signal [40]. As the filter order n increases, the length scale of the fine structure in phase space decreases. For any practical dimension calculation algorithm, if the length scales are small enough, it will be impossible to resolve the fine structure, so the resulting chaotic attractor will appear to have a higher dimension than it actually has.

2.5.5 ▪ Dimension Calculations

The fact that filtering a chaotic signal creates a dimension increase or a complex structure that looks like a dimension increase suggests that a dimension calculation algorithm might be a useful place to start in developing a method to distinguish filters. The method used here was described in [25]. If there is a continuous function between two signals that are embedded in phase space, then if a group of points falls within a small phase space neighborhood on one embedded signal, then the corresponding points on the other embedded signal will also fall within a small neighborhood. We can use this idea to test whether or not a signal has been properly embedded. We start with a signal embedded in d dimensions and ask whether a $d+1''$ dimension is necessary. We pick an index point $v(t_0) = [s(t_0+\tau_1),\ldots,s(t_0+\tau_d)]$ on the d-dimensional signal and find all points within a radius δ of that point, which we can label $v(t_j) = [s(t_j+\tau_1),\ldots,s(t_j+\tau_d)]$. We then look at the points which make up the $d+1''$ component of the vector $v(t_j)$, i.e., the points $v(t_j)_{d+1} = s(t_j+\tau_{d+1})$. We choose a one-dimensional interval ε and ask for the points $v(t_j)$ that lie within the δ radius of the index point $v(t_0)$, how many of the corresponding points $v(t_j)_{d+1}$ lie within ε of $v(t_0)_{d+1}$.

It is possible that some points within the δ ball will fall within the ε interval by chance, so we need to look at a probability distribution. If there are k points in the δ ball, we ask what is the probability p that m of these points will land within the ε ball by chance? We can get this probability from the binomial distribution by assuming that there is a 50% probability that any single point in the δ ball will land within the ε interval.

In the full-dimension calculation algorithm, we start with an initial δ and ε and then reduce ε. The radius δ is then varied until the probability that m of the points from the δ ball land within the ε interval is < 0.05. In this chapter, in order to speed up the calculation, we simply pick an arbitrary value for k, the number of points within δ, and then use the binomial distribution to find the number of points m that have a probability of < 0.05 of landing within the ε radius. In the calculations below, for example, if $k = 20$, then $m = 10$. We take some number of index points $v(t_0) = [s(t_0+\tau_1),\ldots,s(t_0+\tau_d)]$ on the d-dimensional signal and find the 20 nearest neighbors in phase space. Using the indices of these points, the 10 points $s(t_j+\tau_{d+1})$ from this set closest to $s(t_0+\tau_{d+1})$ are found. The one-dimensional Euclidean distance from $s(t_0+\tau_{d+1})$ to the 10th nearest neighbor is ε. The value of ε is averaged over a number of centers to produce $\langle\varepsilon\rangle$.

A large value of $\langle\varepsilon\rangle$ is interpreted as meaning that the $(d+1)$st component of the signal s is not likely to be functionally related to the d dimensional embedding, so the $(d+1)$st component contains information not present in the first d dimensions. The embedding delay τ_{d+1} is varied over some range, and the value of τ_{d+1} that gives the maximum $\langle\varepsilon\rangle$ is taken to be the correct embedding delay for the $(d+1)$st dimension. The maximum value of $\langle\varepsilon\rangle$ is $\langle\varepsilon^*\rangle$ As the embedding dimension d is increased, the curve of $\langle\varepsilon^*\rangle$ vs τ_{d+1} decreases. When d is high enough to give an embedding of s, the curve stops decreasing.

The dimension calculation is similar to [43] in that it tries to determine if there is a continuous function from an embedding in d to an embedding in $d+1$ dimensions. The present calculation includes statistics for estimating confidence in the result.

2.5.6 ▪ Filter Comparison Algorithm

The target identification uses some of the same ideas as the dimension calculation, but it is not a dimension calculation. We begin with a reflection s from a known target. We choose an index point $s(t_0)$ and find the k closest points $s(t_j)$ to the index point. We save the indices t_j of these points as the set $T_j(d=1)$.

We then find the point from s that lags the index point by a delay of τ, or $s(t_0+\tau)$. From the set of points $s(t_j)$ we likewise find $s(t_j+\tau)$. We then find the m points from $s(t_j+\tau)$ that are closest to $s(t_0+\tau)$. Based on concepts from the dimension calculation above, we typically choose $k=20$ and $m=10$. These numbers are chosen large enough so that we get an adequate statistical sampling of the signal, but small enough that the neighbors are confined to local regions of the signal embedded in phase space. The largest distance from the set $|s(t_0+\tau)-s(t_j+\tau)|$ is defined as ε. We then fix τ, choose a new t_0, and repeat the calculation to find a new ε. We continue this process for a number of randomly chosen values of t_0 and average the resulting values of ε to get $\langle\varepsilon\rangle$.

The value of the delay τ will affect the value of $\langle\varepsilon\rangle$, so we next choose a new value of τ and once again average ε over a number of index points on the known signal s to get a new ε. We continue for a number of delay values to get $\langle\varepsilon(\tau)\rangle$, which is $\langle\varepsilon\rangle$ as a function of τ. We choose the maximum value of $\langle\varepsilon(\tau)\rangle$ as $\langle\varepsilon_s^*\rangle$ for dimension $d=1$. We record the corresponding value of τ as $\tau_{d=1}$. We also save the indices t_0 as the set $T_0(d=1)$, and we save the indices t_j of the neighbors corresponding to each t_0 as $T_j(d=1)$. We do not save the actual values of the points, just their indices.

We then embed the known signal s in two dimensions, using the phase space delay $\tau_{d=1}$. We choose a new index point $[s(t_0), s(t_0+\tau_{d=1})]$ and find the k nearest neighbors, $[s(t_j), s(t_j+\tau_{d=1})]$. We choose a different delay τ and from the set of k neighbors we then find the m points $s(t_j+\tau)$ that are closest to $s(t_0+\tau)$. The maximum value of the distance $|s(t_j+\tau)-s(t_0+\tau)|$ is ε. We continue by choosing different index points and delays as in the one-dimensional case to find the maximum $\langle\varepsilon_s^*\rangle$ for $d=2$ and the corresponding $\tau_{d=2}$. Once again, we record the indices t_0 as the set $T_0(d=2)$ and the indices t_j as $T_j(d=2)$.

For the known signal s, we are simply doing the dimension calculation from above, but we are recording the values of t_0, t_j, τ, and $\langle\varepsilon_s^*\rangle$ for each dimension. We continue the calculation process for the known signal s for as many dimensions as are necessary. Typically, if we know the embedding dimension of the signal, we will continue the calculation for two or three higher dimensions.

We now have an unknown signal u that we wish to compare with s. For the signal u, we do not do any phase space searching. For the first dimension, we take an index t_0 from the set $T_0(d=1)$ and the corresponding indices t_j from the set $T_j(d=1)$. As before, for each t_0, there are k possible values of t_j. The points $u(t_j)$ may or may not be close to $u(t_0)$. Then, using the stored value of $\tau_{d=1}$, we find the m members of the set of points $u(t_j+\tau_{d=1})$ that are closest to $u(t_0+\tau_{d=1})$. The maximum of the distance $|u(t_j+\tau_{d=1})-u(t_0+\tau_{d=1})|$ is ε. We next choose other indices t_0 and the corresponding t_j values from the sets $T_0(d=1)$ and $T_j(d=1)$ and average the resulting values of ε to get $\langle \varepsilon_u * \rangle$ for $d=1$.

We continue this procedure for higher dimensions. The result is a set of values $\langle \varepsilon_u * \rangle$ for each dimension. If the signal u is the same as the signal s, then we are just repeating the dimension calculation, and $\langle \varepsilon_u * \rangle = \langle \varepsilon_s^* \rangle$ for each dimension. If u is not the same as s, then for dimensions that are greater than or equal to the embedding dimension of s, $\langle \varepsilon_u * \rangle \gg \langle \varepsilon_s^* \rangle$.

2.5.7 ▪ False Matches

We are assuming in our algorithm that a signal from the same target that generated a reference signal gives the smallest possible phase space radius $\langle \varepsilon^* \rangle$. Is there some other target which could give a smaller value of $\langle \varepsilon^* \rangle$? We can't prove whether or not a false match is possible, but we can provide some arguments that given enough data, a false match is unlikely.

We approximate the target reflection as a sum of delayed versions of the transmitted signal:

$$s_f(i)=\sum_{l=1}^{n} a_l s(t_i-t_l). \tag{2.21}$$

We embed this signal in a phase space with an embedding dimension of d and delays of $\tau_{d=1}, \tau_{d=2}, \ldots, \tau_{d=d}$.

For our reference signal, we find an index point $v(t_0)=[s_f(t_0+\tau_{d=1}),\ldots,s_f(t_0+\tau_{d=d})]$ and k near neighbors $v(t_j)=[s_f(t_j+\tau_{d=1}),\ldots,s_f(t_j+\tau_{d=d})]$. Given these embedded points, we then find a one-dimensional phase space radius ε such that for m of the k near neighbors, $|s_f(t_j+\tau_{d=d+1})-s_f(t_0+\tau_{d=d+1})|\leq\varepsilon$.

We save the indices of all k of the neighbors. We then embed a reflection from an unknown target, z_f. We then ask whether, for m of these k points, $|z_f(t_j+\tau_{d=d+1})-z_f(t_0+\tau_{d=d+1})|\leq\varepsilon$. If this condition holds on average over the entire unknown signal, then we say that the unknown target matches the reference target.

Applying (2.21), the target matching condition becomes

$$\left|\sum_{l=1}^{n} a_l\left[z(t_j-t_l+\tau_{d=d+1})-z(t_0-t_l+\tau_{d=d+1})\right]\right|\leq\varepsilon. \tag{2.22}$$

Is there some other target which could also match the condition of (2.22)? One way for another target to match the condition of (2.22) is for a set of coefficients b_l to exist such that

$$\begin{aligned}&\sum_{i=1}^{n} b_i\left[z(t_j-t_i+\tau_{d=d+1})-z(t_0-t_i+\tau_{d=d+1})\right]\\&=\sum_{l=1}^{n} a_l\left[z(t_j-t_l+\tau_{d=d+1})-z(t_0-t_l+\tau_{d=d+1})\right]\end{aligned} \tag{2.23}$$

so that the phase space neighbors for the target with coefficients b_l are the same as the neighbors for the target with a_l.

For (2.23), if the coefficients a_l and delays t_l are fixed, then the unknowns are the coefficients b_l and the delays t_i. Therefore, for a target of order n, there are $2n$ unknown quantities to be found. To find these unknowns, we get one equation from each phase space neighbor. If the number of neighbors is m and we find the average phase space radius $\langle \varepsilon^* \rangle$ by averaging over M points on the attractor, then we have a total of $n \times M$ equations. For example, if we use 10 neighbors and average over 1000 centers on the embedded signal, there are 10,000 equations. If we have fewer equations than unknowns, then there are an infinite number of solutions, so there are many other targets that can match our reference target. If there are more equations than unknowns, then there is no solution to (2.23). While lack of a solution does not prove that no other target can match our reference target, it at least gives us some confidence that the odds of a false match are small. In the example just quoted, the target would have to appear as 10,000 point scatters. In our testing below, we use signals that have low range resolution (a typical target is < 10 wavelengths in extent), so the number of individual scattering centers should be small.

2.5.8 ▪ Noise Effects

For two noise free points separated by a distance $\delta_0^2 = (z_f(t_i) - z_f(t_j))^2$, the separation with an added noise signal η will be $\delta_1^2 = (z_f(t_i) + \eta(t_i) - z_f(t_j) - \eta(t_j))^2$, or $\delta_1^2 = \delta_0^2 + [\eta(t_i) - \eta(t_j)]^2 + 2[\eta(t_i) - \eta(t_j)][z(t_i) - z(t_j)]$. If the noise and the signal are uncorrelated, then averaging over the signal will make the last term go to 0, so the increase in the mean squared distance between points will be equal to the mean squared difference for the noise signal. As a result, $\langle \varepsilon^* \rangle$ increases by the same amount for all signals, so the smallest value of $\langle \varepsilon^* \rangle$ is still obtained when the reference signal matches the unknown signal.

2.5.9 ▪ Chaotic Signals

To demonstrate the target identification method, we first used the x signal from the Hénon map of (2.19), with α=1.29 and $\beta = 0.3$.

The dimension calculation described above was applied to the x signal. Because this system is a map, the natural delay unit is one iteration step, so the delay times for the phase space embedding did not have to be optimized; instead, for an embedding in d dimensions, the delays were $0, 1, 2, \ldots, d-1$. Figure 2.10 shows the results of a dimension calculation for a time series of 50,000 points of the x variable.

The plot of $\langle \varepsilon^* \rangle$ in Figure 2.10 levels off at $d = 2$, showing that two dimensions are enough to embed the Hénon signal. This result is consistent with the Kaplan–Yorke dimension of 1.3.

2.5.10 ▪ Simple Targets

The fine structure in phase space can be used to differentiate two filters. As in the description above, we find nearby phase space neighbors for an embedded version of the filtered

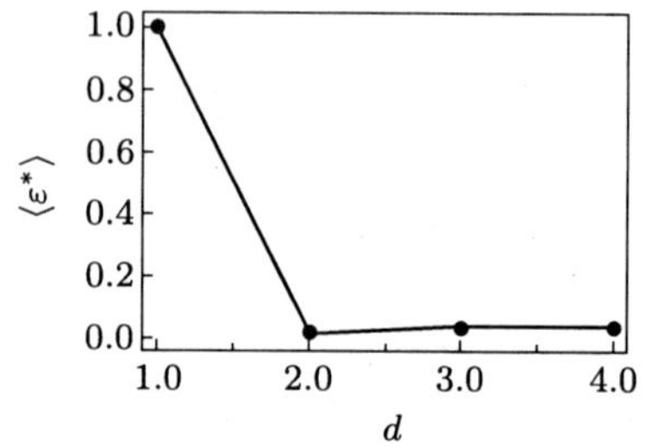

Figure 2.10. *Dimension calculation for x signal from the Hénon attractor.* $\langle \varepsilon^* \rangle$ *is the average phase space radius in the* $d+1''$ *dimension for a group of points in the d dimension.*

signal z. We then create a second filter:

$$u(n) = x(n) + x(n+k+1). \tag{2.24}$$

We use the indices of the phase space neighbors found for z to calculate ε^* for u. Figure 2.11(a) shows $\langle \varepsilon^* \rangle$ for z and u calculated using the indices from z. The constants in (2.19) were $\alpha = 1.29$ and $\beta = 0.3$, resulting in Lyapunov exponents of 0.78 bits and −2.517 bits.

Because the fine scale phase space structure is very sensitive to the filtering, the value of $\langle \varepsilon^* \rangle$ for u is larger than for z. The values of $\langle \varepsilon^* \rangle$ are not appreciably different until the delay k reaches 4; for shorter delays, the filter does not create the fine scale structure in phase space.

The contraction rate for the map determines what delay is necessary before fine scale structure is formed. Figure 2.11(b) shows $\langle \varepsilon^* \rangle$ for z and u for $\alpha = 1.29$ and $\beta = 0.15$, in which case the Lyapunov exponents for the Hénon map were 0.61 bits and −3.3473 bits.

The two-dimensional Hénon map expands (or contracts) areas in phase space at a rate given by the absolute value of the determinant of the Jacobian matrix of the map (2.19).

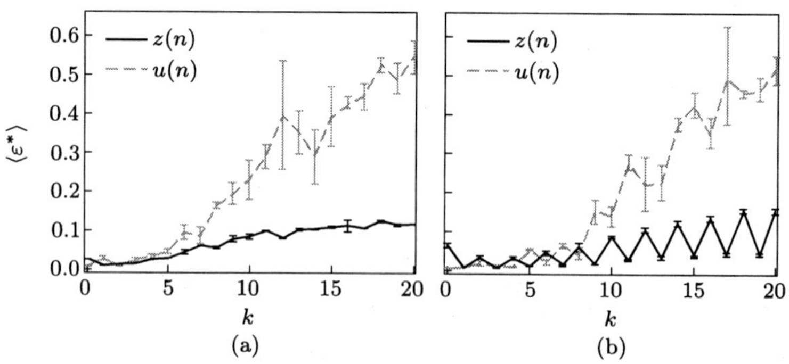

Figure 2.11. $\langle \varepsilon^* \rangle$ *calculated for z and u using indices from z. For* (a)*, the Hénon map parameters from* (2.19) *were* $\alpha = 1.29$ *and* $\beta = 0.3$*. For* (b)*, the Hénon map parameters from* (2.19) *were* $\alpha = 1.29$ *and* $\beta = 0.15$*. The error bars are obtained by dividing the signal into* 10 *equal length sections and averaging the results. The error bars are statistical errors caused by a finite number of samples.*

For the Hénon map then, the expansion rate for areas in phase space is [36]

$$\left\| \begin{matrix} -2\alpha x & 1 \\ \beta & 0 \end{matrix} \right\| = |-\beta|. \tag{2.25}$$

A value of $\beta < 1$ means that areas are contracted. A smaller value of β means more compression per iteration. When $\beta = 0.15$, a filter with a delay of at least eight points was necessary to differentiate two targets, while a delay of four points was necessary when $\beta = 0.3$. As the areas in phase space are compressed at a faster rate, it takes a delay of more points to generate the fine structures in phase space that make it possible to use our algorithm to differentiate filters. Note that the creation of this fine structure is not related to the decrease of the autocorrelation of the x signal over longer and longer delays; when $\beta = 0.3$, the autocorrelation of the x signal dropped to 1/3 of its initial value within two points, but when $ta = 0.15$, it took the autocorrelation 12,000 points to drop by the same amount.

2.5.11 ▪ Finite Difference Time Domain Simulations

We simulated this target identification method with simulated missiles similar to those shown in Figure 2.4. In this case, we simulated three different missiles.

The three missiles were all slightly under 5 m in length with 2 m wingspans. Details varying from model to model included cavity positions, radomes, inlets, wing and tail sweeps, etc. A six-dimensional chaotic map that was linear except for a modulus function was used to generate a chaotic signal $x(n)$ which was frequency modulated onto a sinusoidal carrier (2.17). In this case, the map parameters (Table 2.1) were chosen randomly, but the map Jacobian was evaluated to ensure that the resulting signal was chaotic.

Table 2.1. *Parameters for the chaotic map of* (2.17).

i	$p(i)$
1	−0.53392398
2	−0.40588415
3	0.343772781
4	0.044160236
6	0.443750493

The map signal was modulated onto a sine wave according to (2.18). The waveform had a center frequency of 250 MHz and a 20% bandwidth. Three hundred cycles of this chaos-based waveform were generated and divided into 10 pulses of 30 cycles each. The FDTD code transmitted each pulse as an H-polarization electromagnetic wave at each of the four missiles along the 0° azimuth, 0° elevation line. The monostatic backscattered pulse was collected in each instance and employed in this analysis. Missile modeling grid incremental size of 4 cm was selected to be smaller than 1/20th of the free space wavelength at 275 MHz to ensure an accurate FDTD modeling. The problem size and run times were limited so that no rounding error inaccuracies were evident in the returned waveforms.

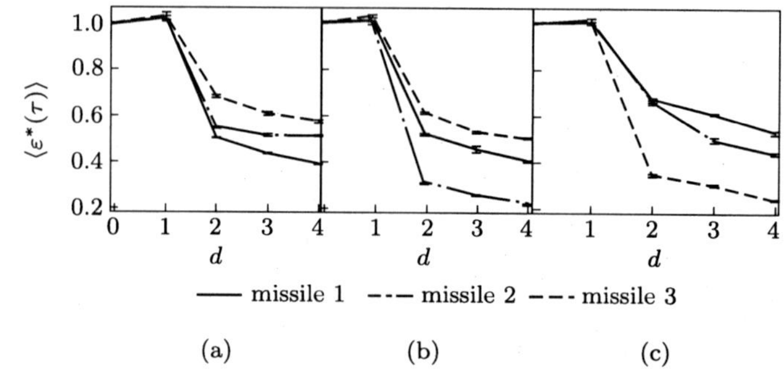

Figure 2.12. (a) *A signal reflected from missile* 1 *is compared to reference files generated from signals reflected from missiles* 1, 2, *and* 3. *The error bars are obtained by dividing the signal into* 10 *equal length sections and averaging the results. The error bars are statistical errors caused by a finite number of samples.* (b) *A signal reflected from missile* 2 *is compared to reference files generated from signals reflected from missiles* 1, 2, *and* 3. (c) *A signal reflected from missile* 3 *is compared to reference files generated from signals reflected from missiles* 1, 2, *and* 3.

In order to apply our target discrimination algorithm, 1800 points from the middle of each reflected pulse were used. The beginning and end of each pulse were dropped to avoid transients and simulation artifacts. Five hundred index points were randomly picked on the first pulse for each data set, and all 10 pulses were simultaneously searched for phase space neighbors. Phase space delays of up to 90 points (6.75 ns) and phase space dimensions up to five were considered.

All three missiles were illuminated from directly ahead. The signatures of the missiles will be different for different aspect angles, but the aspect angle of a moving missile can be deduced from its trajectory.

Figure 2.12 shows $\langle \varepsilon^* \rangle$ values for signals reflected from missiles 1, 2, and 3. Reference files were generated from reflected signals from missiles 1, 2, and 3.

The lowest values for $\langle \varepsilon^* \rangle$ in Figure 2.12 occur when the reflected signal is compared to a reference from the same missile, so the target is correctly identified from the reflected signal.

2.6 ▪ Conclusions

The broad band nature of chaotic signals at first seems to make them obvious choices for radar, but the combination of modern signal design methods and high speed digital to analog conversion makes the use of chaos unnecessary in many situations. The only thing that matters for imaging point targets is the power spectrum of the signal, so filtered random noise signals work just as well. Even for complex targets, there is often no advantage for the phase space structures that are always a characteristic of chaotic signals. The requirement that radar signal have a constant envelope in order to maximize the ratio of average power to peak power also means that some of the unusual properties of chaotic signal are lost.

Phase space analysis based on the properties of chaotic signals may be useful for radar, however. When modulated onto a constant envelope carrier, even random signals are finite dimensional, so they may be analyzed in phase space. While cross correlation (or pulse compression) is the best way to reveal a target location in Gaussian noise, much information carried by the signal reflected from a target is lost during pulse compression. Even if cross correlation is used for signal detection, the signal that is used for correlation may be optimized for locating specific targets.

References

[1] B. Flores, E. A. Solis, and G. Thomas, "Assessment of chaos-based FM signals for range-Doppler imaging," *IEEE Proc. Radar Sonar Navigation*, 150, pp. 313–322, 2003.

[2] L. Fortuna, M. Frasca, and A. Rizzo, "Chaotic pulse position modulation to improve the efficiency of sonar sensors," *IEEE Trans. Instrumentation and Measurement*, 52, pp. 1809–1814, 2003.

[3] Y. Hara, T. Hara, T. Seo, H. Yanagisawa, P. Ratliff, and W. Machowski, "Development of a chaotic signal radar system for vehicular collision-avoidance," *Proceedings of the 2002 IEEE Radar Conference*, pp. 227–232, 2002.

[4] F. Y. Lin and J. M. Liu, "Chaotic radar using nonlinear laser dynamics," *IEEE Journal of Quantum Electronics*, 40, pp. 815–820, 2004.

[5] W. Machowski and P. Ratliff, "Performance indications of a novel chaotic-signal FM-CW radar for multi-user applications," *Proc. 2002 International Radar Conference*, Edinburgh, UK, 2002, pp. 474–477.

[6] V. Venkatasubramanian and H. Leung, "A novel chaos-based high-resolution imaging technique and its application to through-the-wall imaging," *IEEE Signal Processing Letters*, 12, pp. 528–531, 2005.

[7] P. Willett, J. Reinert, and R. Lynch, "LPI waveforms for active sonar?," *2004 IEEE Aerospace Conference Proceedings*, pp. 2236–2248, 2004.

[8] K. A. Lukin, "Noise radar technology," *Telecommunications and Radio Engineering*, 55, pp. 8–16, 2001.

[9] P. R. Atkins and A. J. Fenwick, "Chaotic signals in a multipath channel," *Proceedings of the Institute of Acoustics*, 26, pp. 154–162, 2004.

[10] M. I. Skolnik, *Introduction to Radar Systems*, McGraw-Hill, New York, 2001.

[11] P. M. Woodward, *Probability and Information Theory, with Applications to Radar*, Artech House, Dedham, MA, 1980.

[12] J. S. Bergin, P. M. Techau, J. E. D. Carlos, and J. R. Guerci, "Rader waveform optimization for colored noise mitigation," *IEEE Radar Conference*, 2005.

[13] K. T. Wong, "Adaptive pulse-diverse radar/sonar waveform design," *Proceedings of the 1998 IEEE Radar Conference, RADARCON'98*, pp. 105–110, 1998.

[14] A. V. Oppenheim and R. W. Schafer, *Digital Signal Processing*, Prentice-Hall, Englewood Cliffs, 1975.

[15] J. J. Kroszczynski, "Pulse compression by means of linear-period modulation," *Proc. IEEE*, 57, pp. 1260–1266, 1969.

[16] A. W. Rihaczek and S. J. Hershkowitz, *Theory and Practice of Radar Target Identification*, Artech House, Norwood, MA, 2000.

[17] E. Rothwell, C. Kun-Mu, and D. Nyquist, "Extraction of the natural frequencies of a radar target from a measured response using E-pulse techniques, *IEEE Trans. Antennas and Propagation*, 35, pp. 715–720, 1987.

[18] S. U. Pillai, H. S. Oh, D. C. Youla, and J. R. Guerci, "Optimum transmit receiver design in the presence of signal-dependent interference and channel noise," *IEEE Trans. Inform. Theory*, 46, pp. 577–584, 2000.

[19] R. L. Devaney, *An Introduction to Dynamical Systems*, Benjamin/Cummings, Menlo Park, CA, 1986.

[20] J. M. T. Thompson and H. B. Stewart, *Nonlinear Dynamics and Chaos*, Wiley, New York, 1989.

[21] E. N. Lorenz, "Deterministic non-periodic flow," *Journal of Atmospheric Science*, 20, pp. 130–141, 1963.

[22] F. Takens, *Detecting Strange Attractors in Turbulence*, Springer-Verlag, New York. 1980

[23] T. Sauer, J. A. Yorke, and M. Casdagli, "Embedology," *J. Stat. Phys.*, 65, pp. 579–616, 1991.

[24] H. D. I. Abarbanel, R. Brown, J. J. Sidorowich, and L. S. Tsimring, "The analysis of observed chaotic data in physical systems," *Reviews of Modern Physics*, 65, pp. 1331–1392, 1993.

[25] L. M. Pecora, L. Moniz, J. Nichols, and T. L. Carroll, "A unified approach to attractor reconstruction," *Chaos*, 17, 013110, 2007.

[26] J.-P. Eckmann and D. Ruelle, "Ergodic theory of chaos and strange attractors," *Reviews of Modern Physics*, 57, pp. 617–656, 1985.

[27] T. L. Carroll, "Optimizing chaos-based signals for complex radar targets," *Chaos*, 17, 033103, 2007.

[28] T. L. Carroll, "Adaptive chaotic maps for identification of complex targets," *IET Radar Sonar Navigation*, 2, pp. 256–262, 2008.

[29] J. L. Stewart and E. C. Westerfield, "A theory of active sonar detection," *Proc. IRE*, 47, pp. 872–881, 1959.

[30] I. Procaccia, "Universal properties of dynamically complex systems: The organization of chaos," *Nature*, 333, pp. 618–623, 1988.

[31] P. Cvitanovic, G. H. Gunaratne, and I. Procaccia, "Topological and metric properties of Hénon-type strange attractors," *Phys. Rev. A*, 38, pp. 1503–1520, 1988.

[32] J. S. Iwanski and E. Bradley, "Recurrence plots of experimental data: To embed or not to embed," *Chaos*, 8, pp. 861–871, 1998.

[33] F. J. Rachford and T. L. Carroll, *A Chaos Based Waveform Approach to Radar Target Identification*, PIERS Proceedings, Cambridge, MA, pp. 719–722, 2010.

[34] W. H. Press, B. P. Flannery, S. A. Teukolsky, and W. T. Vetterling, *Numerical Recipes*, Cambridge University Press, New York, 1990.

[35] S. A. Nene and S. K. Nayar, "A simple algorithm for nearest neighbor search in high dimensions," *IEEE Transactions on Pattern Analysis and Machine Intelligence*, 19, pp. 989–1003, 1997.

[36] P. Berge, Y. Pomeau, and C. Vidal, *Order within Chaos*, Wiley, New York, 1984.

[37] R. Badii, G. Broggi, B. Derighetti, M. Ravini, S. Ciliberto, and A. Politi, "Dimension increase in filtered chaotic signals," *Phys. Rev. Lett.*, 60, p. 979, 1988.

[38] J. D. Farmer, E. Ott, and J. A. Yorke, "The dimension of chaotic attractors," *Phys. D*, 7, pp. 153–180, 1983.

[39] J.-P. Eckmann, S. O. Kamphorst, D. Ruelle, and S. Ciliberto, "Liapunov exponents from time series," *Phys. Rev. A*, 34, pp. 4971–4979, 1986.

[40] D. S. Broomhead, J. P. Huke, and M. R. Muldoon, "Linear filters and non-linear systems," *J. Roy. Statist. Soc. Ser. B*, 54, pp. 373–382, 1992.

[41] H. Kantz and E. Olbrich, "Scalar observations from a class of high-dimensional chaotic systems: Limitations of the time delay embedding," *Chaos*, 7, pp. 423–429, 1997.

[42] J. Bednar, "On the approximation of FIR by IIR digital filters," *IEEE Transactions on Acoustics, Speech and Signal Processing*, 31, pp. 28–34, 1983.

[43] M. B. Kennel, R. Brown, and H. D. I. A. Abarbanel, "Determing embedding dimension for phase-space reconstruction using a geometrical construction," *Phys. Rev. A*, 45, pp. 3403–3411, 1992.

Chapter 3

Communicating with Exactly Solvable Chaos

Ned J. Corron, Jonathan N. Blakely, and Shawn D. Pethel

Abstract: Recent developments that enable the practical and beneficial use of chaos communications by control of symbolic dynamics are summarized. The key advance is the discovery of a new class of low-dimensional chaotic oscillators that enable simple encoding and coherent reception. These remarkable oscillators are provably chaotic, exhibit a symbolic dynamics with a generating partition, and admit an exact analytic solution as a linear convolution of symbols with a fixed basis function. For encoding information, the exact solution provides an analytic coding function that enables straightforward control of the system's symbolic dynamics. For reception, the existence of a fixed basis function enables a simple matched filter receiver for coherently detecting symbols in the chaotic waveform. System performance exceeds other proposed chaos communications methods and approaches the theoretical limit of binary phase-shift keying (BPSK). Significantly this approach to communication exploits chaos to yield advantages unavailable to conventional modulation of stable, linear waveforms for high-bandwidth data communications.

3.1 ▪ Introduction

In the early 1990s, the nearly simultaneous discoveries of chaos synchronization [1] and control [2] suggested that chaos might be useful for engineering purposes. These initial reports stirred great excitement, and it is no understatement to say that an explosion of research literature followed.

In particular, the discovery of synchronization sparked a torrent of interest in communication using chaotic waveforms. Although synchronization of coupled chaotic oscillators had been previously reported [3, 4], it was the landmark paper of Pecora and Carroll in 1990 that finally drew attention to the phenomenon by suggesting possible application to communications [1]. Subsequently, there has been considerable research and development toward a viable chaos communications technology—both synchronization based and not—and numerous texts have now appeared to document the field [5, 6, 7, 8].

Among the early chaos communications papers was a particularly elegant concept due to Hayes et al. [9, 10], in which information is encoded in the symbolic dynamics of a chaotic oscillator using chaos control. Originally, chaos control was used only to stabilize unstable periodic orbits and steady states in a chaotic oscillator. However, the Hayes approach extended chaos control to target aperiodic trajectories that could encode the information of a message signal. In effect, the random and unpredictable entropy of chaos is thereby replaced with specific information of the prescribed message. Importantly, the assignment of trajectories to message information is made using a symbolic dynamics, in which continuous state values are partitioned using a small number of discrete labels [11, 12]. Practically, the enabling element of the Hayes approach is that it exploits the shift dynamic of a symbolic representation to efficiently encode a continuous stream of message data using consistently small control perturbations [13].

This elegant encoding method, which truly exploits the properties of chaos in an efficient and meaningful way, has unfortunately suffered from the lack of an equally elegant method for decoding the information. That is, a complementary receiver that also exploits the properties of chaos to recover the information encoded in the waveform was not immediately recognized. In conventional communications, optimal receivers generally rely on a linear correlation to suppress noise and interference. The best example is a matched filter receiver, which provides optimal detection in the presence of additive white Gaussian noise (AWGN) [14]. In practice, matched filters perform remarkably well in realistic environments and have been a cornerstone of communications for decades.

Unfortunately, chaos is not usually amenable to implementing simple matched filters. Chaos communication methods based on linear correlation receivers, such as chaos shift-keying [15], do not exploit the chaotic nature of the waveforms; instead, the fundamental theory underlying these methods merely tolerates the chaos and performs just as well (or perhaps better) using nonchaotic waveforms. It is possible that synchronization may enable coherent detection, but maintaining high-quality synchronization is difficult in the presence of significant noise [16, 17]. Nonlinear noise reduction approaches are often computationally complex and yield only modest results [18, 19].

However, a new class of low-dimensional chaotic oscillators was recently reported that enables simple and effective coherent reception [20, 21]. These oscillators are remarkable in that they admit an exact analytic solution, are provably chaotic, and exhibit a symbolic dynamics with a generating partition. Importantly, the solution can be written as a linear convolution of the symbolic dynamics and a fixed basis function. The existence of a fixed basis function within the chaotic waveform enables the realization of a simple matched filter receiver for detecting symbols in the chaotic waveform. Furthermore, implementations of these oscillators in electronic circuits have also been reported, suggesting that practical data communications can be realized.

Since they are low-dimensional and exhibit a convenient symbolic dynamics, exactly solvable oscillators are well suited for encoding information using the Hayes approach. In fact, control is particularly effective for these oscillators, since the analytic solution enables exact calculation of control targets necessary for encoding the symbolic dynamics. However, the real advantage is that these exactly solvable oscillators enable the use of a matched receiver to form a complete, practical, and viable communications system. The analytic solutions enable theoretical analysis of the performance of such a system, which is found to be close to the theoretical limit of binary phase-shift keying (BPSK). Thus, it is finally possible to realize an effective chaos communications receiver that exploits the dynamics of chaos and matches the elegance of the Hayes encoding.

Significantly a communication system based on this approach exploits chaos to yield advantages not available to conventional modulation of stable, linear waveforms. First, the efficiency of small amplitude control provides an encoding scheme without modulation transients. As a result, the entire waveform can be used to carry information, and there is no degradation of the correlation receiver due to switching transients. Second, since the modulated waveform stays within the chaotic attractor, channel equalization can be optimized for the transmission of the chaotic waveform and is independent of the message signal. Third, since the data rate can be as high as the topological entropy of the chaotic oscillator, very high percent information bandwidth is available in these systems. As such, the Hayes approach to chaos communications may be ideally suited for very high bandwidth data communications.

This chapter provides the details of the exactly solvable chaotic oscillators and their implication for communication with chaos. Current research is aimed at making this vision of chaos communications practical. Key areas left to be explored include radio-frequency oscillator implementations, high-speed chaos control, data encoding within a restricted symbolic grammar, and error avoidance and correction. Most importantly, future research will need to identify a niche application that can drive and support the development of this novel technology.

3.2 ▪ Communications

The overall goal of any communication system is to convey a message from a source to destination [22]. To this end, a communication system includes a transmitter, channel, and receiver, as shown in Figure 3.1. The purpose of the transmitter is to encode a message in a form suitable for transmission through the channel. In the channel, the transmitted signal may be altered in ways that may or may not be known. Even if known, the distortions may not be reversible. The function of the receiver is then to extract the signal from the channel and recover the message. Any distortions of the signal in the channel can compromise the ability of the receiver to correctly recover the message. It is the objective of communication system design to minimize communication errors and maximize throughput via intelligent choices in transmitter and receiver design.

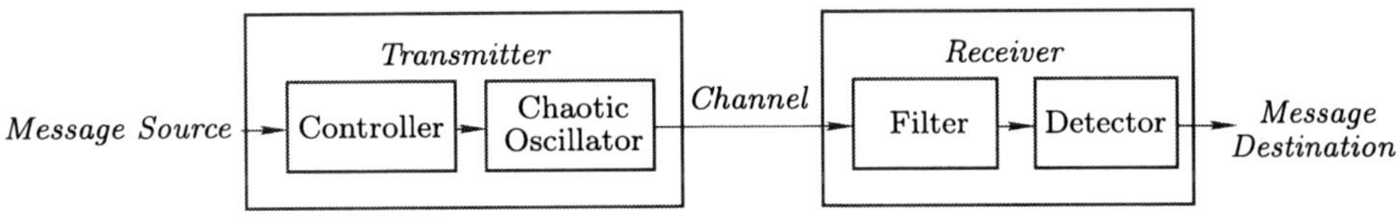

Figure 3.1. *General chaos communications system architecture.*

For the chaos communications considered here, we begin by making the arbitrary choice that the signal transmitted through the channel is a chaotic waveform. Tacitly, we are assuming there is a channel for which a chaotic waveform is justified. It is often assumed that a communication channel is a wireless, through-the-air, radio-frequency link, which is typically characterized by significant noise, interference, multipath, and spectral constraints. However, other scenarios for which chaos may be considered include wired connections, optical fibers, laser systems, and FM radio channels.

As shown in Figure 3.1, a key element of the chaos communication transmitter is a chaotic oscillator. A signal derived from this oscillator is put into the channel for transmission. Information is encoded on the chaotic waveform by a controller, which is the second important element of the transmitter highlighted in the figure. The purpose of the controller is to select and stabilize a particular trajectory within the chaotic attractor to represent the input message signal. Here we advocate the Hayes encoding method, whereby information is encoded in the symbolic dynamics of a low-dimensional chaotic oscillator using small perturbation control [9].

In Figure 3.1, the receiver is shown to contain two parts: a filter and a detector. The purpose of the filter is to accept the received signal from the channel and perform operations designed to reveal the encoded information signal. Part of this filtering intends to undo any known distortions applied by the channel, to the extent such distortions are reversible. However, filtering also serves to minimize the impact of unknown or irreversible distortions that introduce uncertainty in the true value of the transmitted signal. The detector works on the output of the filter and estimates the message signal that was transmitted. Its function is to identify and select a signal that is, in some sense, most likely the encoded message. This selected message is presented to the destination as the decoded signal from the communication system.

In this chapter, we consider the implications of using an exactly solvable chaotic oscillator for communications. In the transmitter such oscillators yield an exact symbolic dynamics for encoding the message information. The controller is facilitated by the analytic representation of control points necessary for targeting a desired trajectory. In the receiver, the waveform's linear representation containing a single basis pulse enables a simple matched filter. Analytic waveform representations enable derivation of optimal detectors and resulting performance metrics. In the end, it is predicted that the theoretical performance of such a chaos communication system can approach or match the theoretical limit of BPSK for an AWGN channel.

3.3 ▪ Exactly Solvable Chaos

In this section, we present two examples of a new class of low-dimensional exactly solvable chaotic oscillators. The first oscillator is characterized by an embedded shift map, and the attractor is topologically similar to the famous Lorenz oscillator [23]. The second contains a tent map, and the attractor includes folded band dynamics as in Rossler's oscillator [24]. Both of these oscillators can be derived from a single model, which was originally proposed as an analytic model of manifold piecewise-linear chaos by Saito and Fujita [25].

It is commonly assumed that the complexity of chaos denies analytic solution, but this is not necessarily true [26, 27]. For difference equations, it has been shown that trajectories of the shift map and baker's map can be written as the convolution of a random process and a basis pulse [28]. Continuous-time chaotic waveforms can be directly constructed using linear superposition without an explicit dynamical system [29, 30].

Chaotic waveforms have been synthesized in reverse time using a randomly driven linear filter [31, 32]. Recently, an exactly solvable chaotic differential equation was described [33].

Each of the two oscillators considered in this section is a hybrid dynamical system that admits an exact analytic solution and a convenient symbolic dynamics. Importantly, each solution can be written as a linear convolution of binary symbols with a fixed basis function. As we argue in the following sections, we find these attributes make each oscillator extraordinarily well suited for application in a chaos communication system using the Hayes encoding.

3.3.1 ▪ Exact Shift Oscillator

For the first oscillator, we consider the manifold piecewise-linear system that exhibits Lorenz-like chaos [25], which was also previously used for circuit demonstrations of chaos control [34]. This hybrid dynamical system contains a scalar state $u(t) \in \Re$ and a discrete state $s(t) \in \{-1, +1\}$. The continuous-time dynamics are described by the differential equation

$$\frac{d^2u}{dt^2} - 2\beta\frac{du}{dt} + (\omega^2 + \beta^2)\cdot(u - s) = 0, \tag{3.1}$$

where $\omega = 2\pi$ and $0 < \beta \leq \ln 2$. Transitions in the discrete state are defined by the guard condition

$$\frac{du}{dt}(t) = 0 \quad \Rightarrow \quad s(t) = \mathrm{sgn}(u(t)), \tag{3.2}$$

meaning that $s(t)$ is set to the sign of $u(t)$ whenever its time derivative vanishes, and $s(t)$ maintains this value until the guard condition is next met. Here, we use the right-continuous signum function

$$\mathrm{sgn}(u) = \begin{cases} +1, & u \geq 0, \\ -1, & u < 0, \end{cases} \tag{3.3}$$

so that the discrete state takes the values $s(t) = \pm 1$. The hybrid system admits two fixed points $u(t) = s(t) = \pm 1$, both of which are unstable.

Figure 3.2 shows a typical solution obtained via numerical integration of the system (3.1)–(3.2) for $\beta = \ln 2$. In this figure, the state $u(t)$ is the continuous waveform, while $s(t)$ appears as the random square wave. Figure 3.3 shows the corresponding phase-space projection. Figure 3.4 shows an alternate projection using time-delay embedding coordinates. The solution waveform exhibits growing harmonic oscillations combined with random-like switching, and a dual-lobe structure is evident in the phase-space projections.

To better understand how this oscillator works, we consider the typical trajectory shown in Figure 3.5. The dynamics of the continuous state $u(t)$ are described by the harmonic oscillator in equation (3.1), which is unstable with $\beta > 0$. Between switching events, the continuous waveform $u(t)$ exhibits growing oscillations about the fixed level $s(t)$. This growth provides the exponential divergence of initial conditions necessary for chaos. Possible switching events occur whenever the derivative of $u(t)$ is zero, which are indicated by dots in Figure 3.5. The discrete state $s(t)$ switches only if the continuous state has crossed a threshold at zero and is of opposite sign to the set point. After a switch, the unstable oscillations grow about the new set point, and the process repeats.

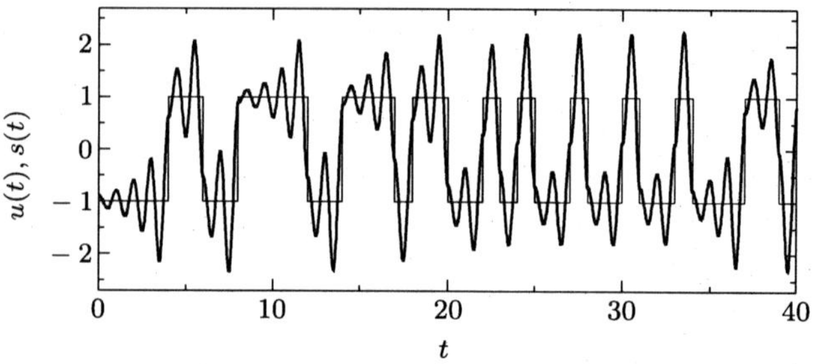

Figure 3.2. *Typical continuous waveform $u(t)$ and discrete state $s(t)$ from numerical integration of the exact shift oscillator for $\beta = \ln 2$.*

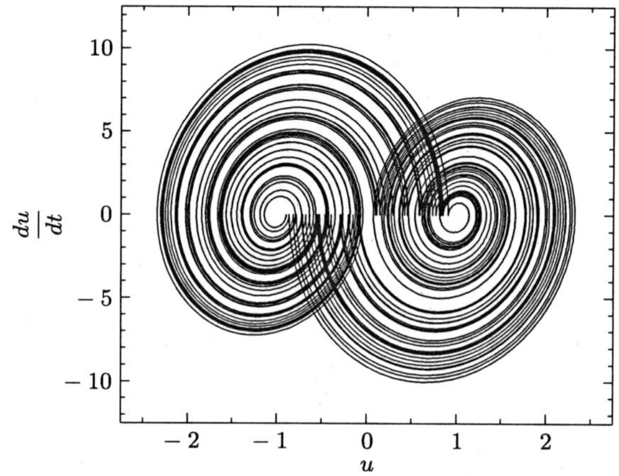

Figure 3.3. *Phase space projection for a typical continuous waveform in the exact shift oscillator with $\beta = \ln 2$.*

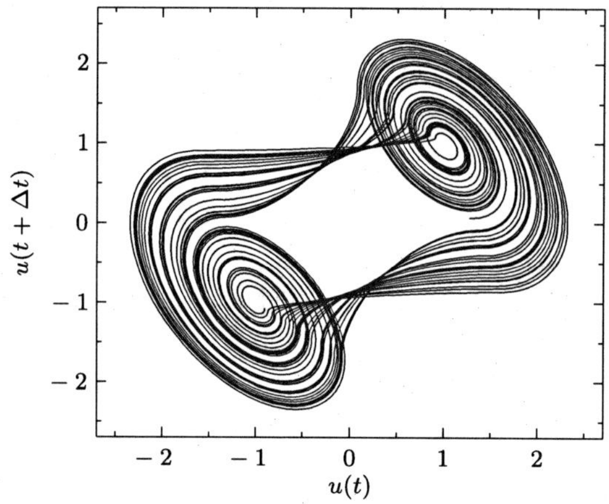

Figure 3.4. *Phase space projection using delay embedding ($\Delta t = 1/3$) for a typical continuous waveform in the exact shift oscillator with $\beta = \ln 2$.*

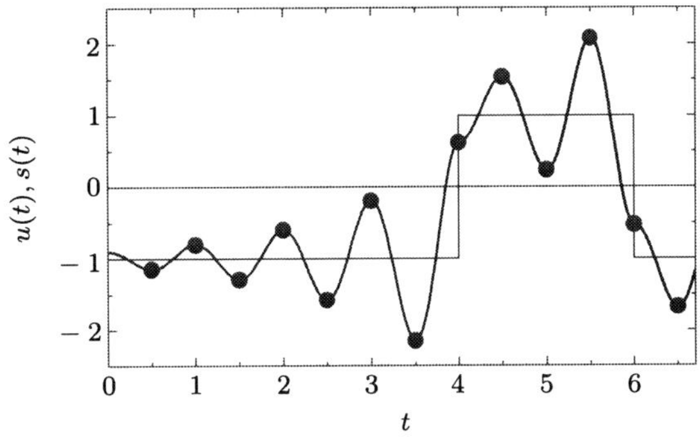

Figure 3.5. *Exponential growth in the harmonic oscillations $u(t)$ producing transitions in the discrete state $s(t)$ when critical points (dots) cross the threshold at zero.*

The numerically generated waveforms and phase space projections give the appearance of chaotic dynamics; in fact, it is straightforward to show the oscillator is chaotic. The nonlinear switching provides a folding action that is essential for chaos. For $\beta \leq \ln 2$, the continuous state at switching times is guaranteed to stay between the set points; thus, oscillations are bounded. Since the differential equation is linear with constant coefficients except at points of instantaneous switching, it follows that $\beta > 0$ is a positive Lyapunov exponent for any trajectory. Thus, the bounded oscillations are unstable and chaotic.

Recently, it was shown that an exact, analytic solution for this hybrid system can be found [20]. Without loss of generality, we consider an initial condition at time $t = t_n$ such that $u(t_n) = u_n$, $\frac{du}{dt}(t_n) = 0$, and $s(t_n) = s_n$, where $s_n = \operatorname{sgn}(u_n)$. That is, we consider an initial state that coincides with a triggering of the guard condition. The solution to the initial value problem is then given by the linear convolution

$$\begin{aligned} u(t) &= \sum_{m=0}^{\infty} s_m \cdot P(t - t_n - m), \\ s(t) &= \sum_{m=0}^{\infty} s_m \cdot \phi(t - t_n - m), \end{aligned} \tag{3.4}$$

where $t \geq 0$ and each symbol $s_m = \pm 1$. The fixed basis functions are

$$P(t) = \begin{cases} \left(1 - e^{-\beta}\right) e^{\beta t} \left(\cos \omega t - \dfrac{\beta}{\omega} \sin \omega t\right), & t < 0, \\ 1 - e^{\beta(t-1)} \left(\cos \omega t - \dfrac{\beta}{\omega} \sin \omega t\right), & 0 \leq t < 1, \\ 0, & 1 \leq t, \end{cases} \tag{3.5}$$

and

$$\varphi(t) = \begin{cases} 0, & t < 0, \\ 1, & 0 \leq t < 1, \\ 0, & 1 \leq t, \end{cases} \tag{3.6}$$

which are shown in Figure 3.6 for $\beta = \ln 2$. In this solution, each binary symbol s_m modulates the fixed basis functions centered at time $t = t_n + m$. Thus, it is right to think

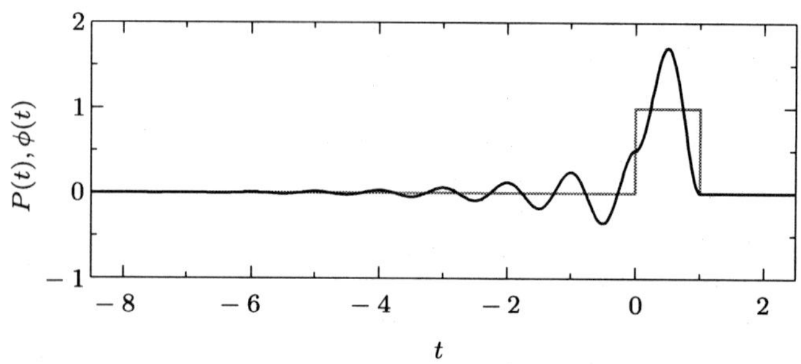

Figure 3.6. *Basis function $P(t)$ for the exact shift oscillator with $\beta = \ln 2$.*

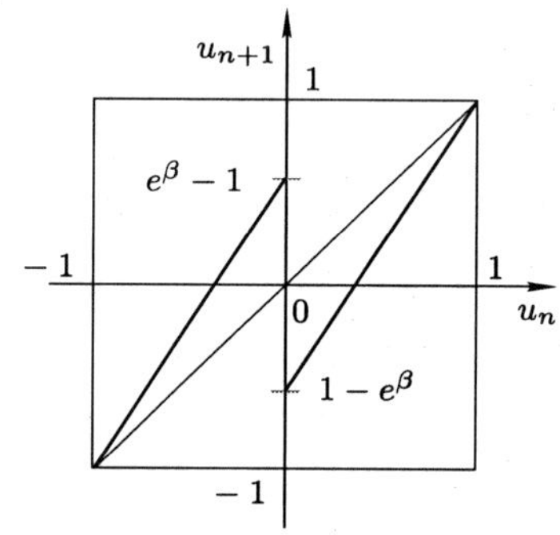

Figure 3.7. *Return map for continuous state in the exact shift oscillator sampled for each symbol.*

of the symbol s_m as the information emitted by the oscillator at time $t = t_n + m$, and that the oscillator emits one symbol with each unit of time. We note that the solution reveals that the oscillator's state only depends on the current and future symbols.

As evident in the solution (3.4), an infinite sequence of future symbols completely and uniquely specifies a solution to the hybrid system. Thus, the future symbols also serve to label each possible solution. To understand the set of possible solutions, we consider a return map derived from the continuous waveform sampled for each symbol. Evaluating the solution at the initial time yields

$$u_n = (1 - e^{-\beta}) \sum_{m=0}^{\infty} s_{m+n} e^{-m\beta}. \tag{3.7}$$

Then the first return satisfies the recurrence relation

$$u_{n+1} = e^{\beta} u_n - (e^{\beta} - 1) \cdot \operatorname{sgn}(u_n), \tag{3.8}$$

which also coincides with a triggering of the guard condition (3.2). Since the oscillator is autonomous, this map yields an iterated one-dimensional map for successive returns sampled at unit times. As shown in Figure 3.7, the return map for the continuous state is piecewise linear and characterized by the slope $e^{\beta} > 1$. As such, the iterated map is a chaotic shift, with Lyapunov exponent $\lambda = \beta$. Furthermore, the symbols $s_n = \operatorname{sgn}(u_n)$ form a symbolic dynamics for the map, with a generating partition at $u_n = 0$. For $\beta = \ln 2$, the map is a full shift and the grammar is unrestricted; however, for $\beta < \ln 2$, the grammar is restricted and not every sequence of symbols is allowed.

The implication is that the hybrid dynamical system is also chaotic. Indeed, the form of the solution (3.4) yields

$$\begin{aligned} u(t+1) &= \sum_{m=0}^{\infty} s_{m+1} \cdot P(t-t_n-m), \\ s(t+1) &= \sum_{m=0}^{\infty} s_{m+1} \cdot \varphi(t-t_n-m), \end{aligned} \tag{3.9}$$

revealing a shift dynamics also exists in the continuous dynamics. Consequently, the symbols also form a symbolic dynamics for the hybrid oscillator.

Importantly, the function (3.7) provides an initial condition for the oscillator to generate a waveform exhibiting a given future symbol sequence. Such a relationship is critical for implementing chaos communications using control of symbolic dynamics, where it is used to calculate the requisite control points in targeting a desired symbolic trajectory. In this context, we recognize (3.7) as the inverse coding function for this oscillator [9].

Since the solution is a linear convolution, the spectral properties of the chaotic continuous waveform can be inferred from the Fourier transform of the basis function. It is

$$p(\Omega) = \int_{-\infty}^{\infty} P(t) e^{i\Omega t} dt = \frac{e^{i\Omega} - 1}{i\Omega} \cdot \frac{\omega^2 + \beta^2}{\omega^2 + (\beta + i\Omega)^2}, \tag{3.10}$$

where $i = \sqrt{-1}$. Thus, the power spectrum for the chaotic waveform exhibits zeros at the bit rate and its harmonics, similar to a conventional digital communications waveform [14]. The power spectrum for $\beta = \ln 2$ is plotted in Figure 3.8.

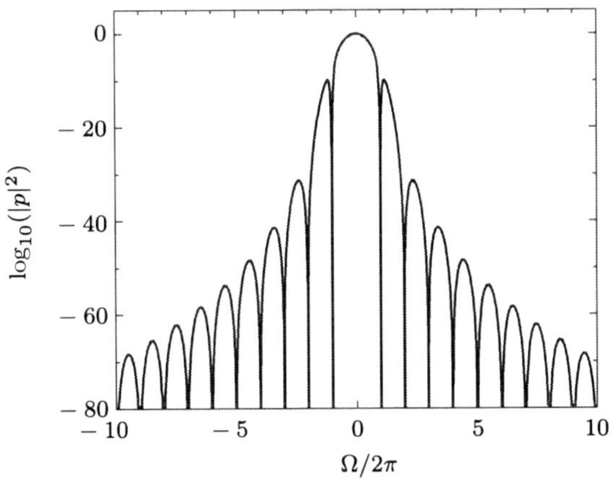

Figure 3.8. *Power spectrum $p(\Omega)$ of the basis pulse for the exact shift oscillator with $\beta = \ln 2$.*

A circuit schematic depicting an implementation of the exact shift oscillator is shown in Figure 3.9. This hybrid circuit includes both analog and digital components. Analog circuitry realizes the continuous state $u(t)$ as the tank voltage in a harmonic LC oscillator in parallel with a negative resistor that supplies gain. The digital circuitry realizes the dynamics of the discrete switching state $s(t)$. The circuit does not use an explicit implementation of the guard condition such as shown in a previous report [20]. Instead, the circuit achieves equivalent operation using two threshold comparators to detect the sign of the tank voltage $u(t)$ and its derivative, which are combined in a simple logic circuit to generate the binary logic signal $s(t)$. The simplicity of the circuit suggests that high-frequency implementations of the exact shift oscillator may be easily realized.

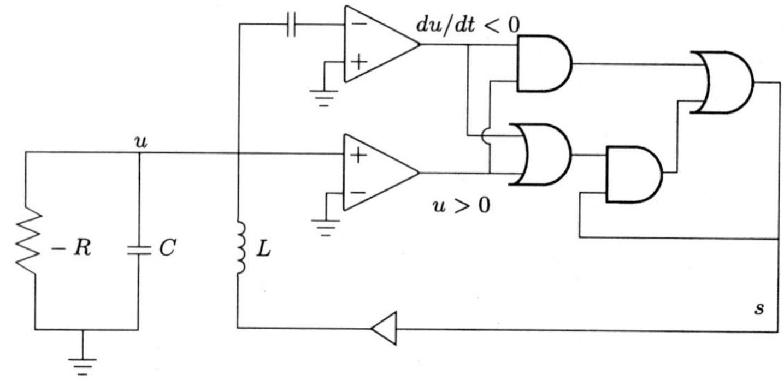

Figure 3.9. *Circuit realization of exact shift oscillator.*

3.3.2 ▪ Exact Folded-Band Oscillator

For the second exactly solvable oscillator, we consider the manifold piecewise-linear system that exhibits folded-band chaos [25]. This system is also a hybrid dynamical system, containing a continuous state $u(t) \in \mathfrak{R}$ and a discrete state $s(t) \in \{0,1\}$. The continuous-time dynamics are again described by the differential equation

$$\frac{d^2u}{dt^2} - 2\beta\frac{du}{dt} + (\omega^2+\beta^2)\cdot(u-s) = 0, \tag{3.11}$$

where $\omega = 2\pi$ and $\beta > 0$. However, the discrete dynamics are now defined by the guard condition

$$\frac{du}{dt}(t) = 0 \quad \Rightarrow \quad s(t) = H(u(t)-1), \tag{3.12}$$

where

$$H(x) = \begin{cases} 1, & x > 0, \\ 0, & x \leq 0, \end{cases} \tag{3.13}$$

is the left-continuous Heaviside function. We note this hybrid system admits the unique fixed point $u(t) = s(t) = 0$, which is unstable.

Typical waveforms calculated numerically for this hybrid system with $\beta = 0.81 \cdot \ln 2$ are shown in Figure 3.10. In this figure, the continuous state shows intervals of various duration in which oscillations grow about the fixed state $s(t) = 0$. Each of these intervals concludes with a switching event where $s(t) = 1$ for one half time unit, during which the growing oscillation is squelched.

The operation of the oscillator is better explained using the short segment shown in Figure 3.11. Initially the unstable linear oscillator (3.11) generates a growing oscillation about the fixed set point $s(t) = 0$. During these oscillations, the guard condition (3.12) is triggered at each critical point, indicated by dots in the figure. While oscillations are small

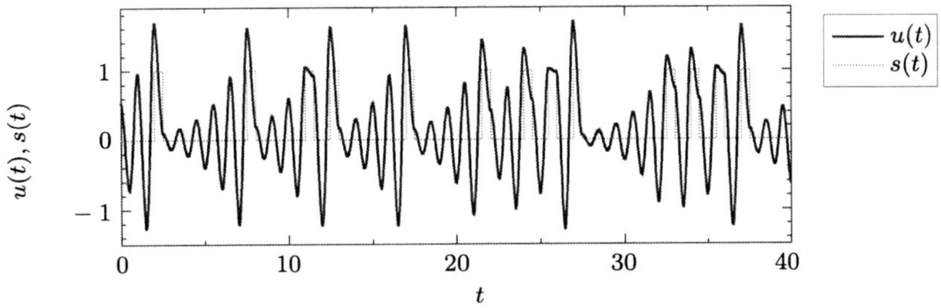

Figure 3.10. *Typical waveforms calculated numerically for the exact folded-band oscillator with $\beta = 0.81 \cdot \ln 2$. Figure courtesy of US Army.*

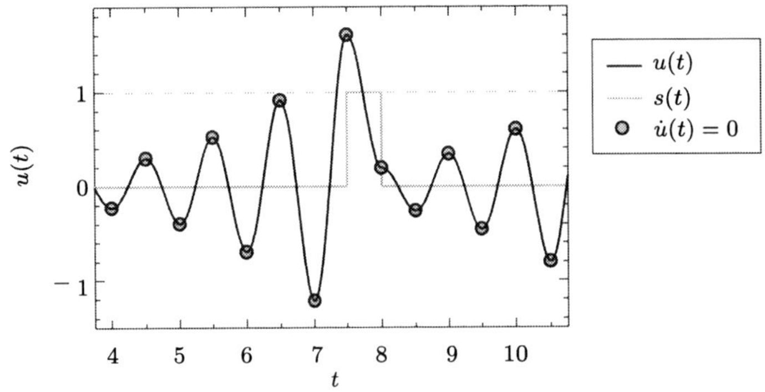

Figure 3.11. *Waveform segment illustrating the hybrid dynamics in the exact folded-band oscillator. Figure courtesy of US Army.*

(amplitude less than one), the discrete state remains unchanged. However, the growing oscillations eventually yield a guard condition with $u(t) > 1$, triggering a transition to $s(t) = 1$. The continuous state now starts to oscillate about the elevated set point. At the next guard condition, which occurs a half cycle later, the state $u(t) < 1$ is again small, resetting $s(t) = 0$ and enabling the growing oscillation process to repeat.

A phase space projection of the continuous state is shown in Figure 3.12. In this projection the solution trajectory spirals outward from the origin, rotating in a clockwise direction with time. Switching events triggered at the guard condition are evident on the positive u axis, folding the outer trajectories back into the center of the spiral. An alternate phase space projection using time-delay embedding is shown in Figure 3.13. Here the trajectory again spirals outward in a clockwise direction, but now the switching events are lifted off the spiral and the folding action is more clearly visible. In these projections, the oscillations appear chaotic. Similar to the exact shift oscillator, the continuous dynamics are described by the linear differential equation (3.11), so $\beta > 0$ defines a positive Lyapunov exponent for any trajectory. Consequently, any bounded oscillations are also chaotic.

As with the exact shift oscillator, the exact folded-band oscillator also admits an exact analytic solution [21]. Without loss of generality, we again consider an initial condition corresponding to a guard condition at time $t = t_n$ such that $u(t_n) = u_n$, $\frac{du}{dt}(t_n) = 0$, and

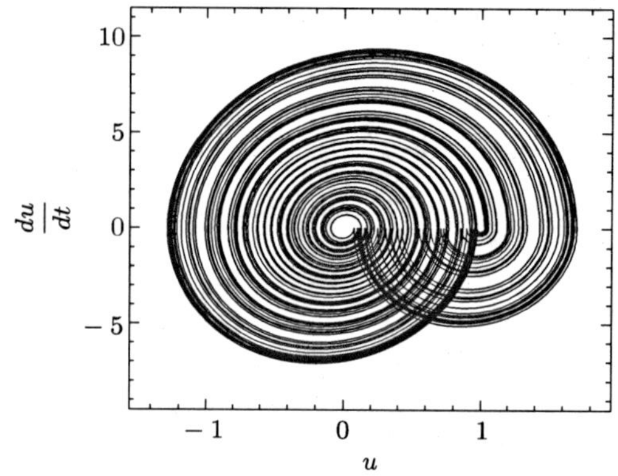

Figure 3.12. *Phase space projection of a typical waveform generated by the continuous state in the exact folded-band oscillator with* $\beta = 0.81 \cdot \ln 2$. *Figure courtesy of US Army.*

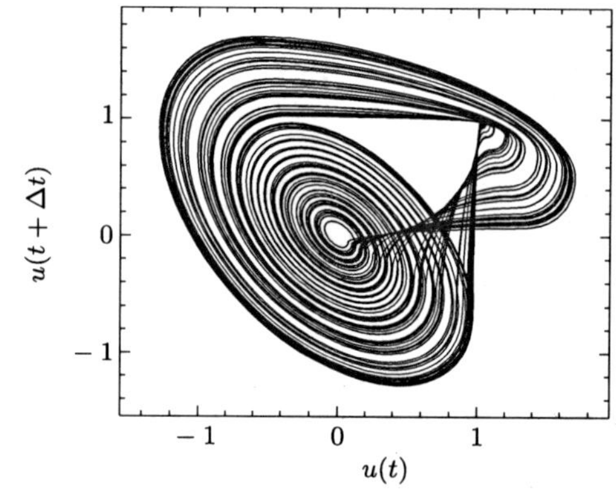

Figure 3.13. *Phase space projection using delay embedding* ($\Delta t = 1/3$) *for a typical continuous waveform generated by the continuous state in the exact folded-band oscillator with* $\beta = 0.81 \cdot \ln 2$. *Figure courtesy of US Army.*

$s(t_n) = s_n$, where $s_n = H(u_n - 1)$. The solution can be written as the linear convolution

$$\begin{aligned} u(t) &= \sum_{m=0}^{\infty} \sigma_m \cdot Q\left(t - t_n - \frac{m}{2}\right), \\ s(t) &= \sum_{m=0}^{\infty} \sigma_m \cdot \psi\left(t - t_n - \frac{m}{2}\right), \end{aligned} \tag{3.14}$$

where

$$Q(t) = \begin{cases} (1 + e^{-\beta/2}) e^{\beta t} \left(\cos \omega t - \dfrac{\beta}{\omega} \sin \omega t\right), & t < 0, \\ 1 + e^{\beta(t-1/2)} \left(\cos \omega t - \frac{\beta}{\omega} \sin \omega t\right), & 0 \le t < \dfrac{1}{2}, \\ 0, & \dfrac{1}{2} \le t, \end{cases} \tag{3.15}$$

and

$$\psi(t) = \begin{cases} 1, & 0 \le t < 1/2, \\ 0 & \text{otherwise,} \end{cases} \tag{3.16}$$

are fixed basis functions for the continuous and discrete states, respectively. The basis functions are plotted in Figure 3.14. Each coefficient σ_m is an amplitude that weights the basis functions at the corresponding time lag in the convolution. In particular, the discrete state simply yields $s(t) = \sigma_m$ for $\frac{m}{2} \le t - t_n < \frac{m+1}{2}$; hence, the amplitude values are restricted to the set $\sigma_m \in \{0, 1\}$. This general solution is valid for $t \ge t_n$.

At the initial time $t = t_n$, the solution evaluates to give

$$u_n = (1 + e^{-\beta/2}) \cdot \sum_{m=0}^{\infty} \sigma_m (-e^{-\beta/2})^m \tag{3.17}$$

and $s_n = \sigma_0$. Thus, for a given initial condition, it is possible to find a sequence of amplitudes $\sigma_m \in \{0, 1\}$ that satisfy (3.17), but the sequence may not be unique. The correct sequence (i.e., the sequence exhibited by the deterministic dynamics) also satisfies a (nontrivial) grammar restriction. In fact, this problem is related to the unique representation of real numbers using noninteger bases [35]. Conversely, (3.17) provides an initial condition that generates a given future sequence for the binary state $s(t)$. Such a relationship is important for chaos communications using symbolic dynamics control, where it provides the requisite control points for a targeted symbolic trajectory. In this context, we recognize (3.17) as the inverse coding function for this oscillator [9].

To analyze the dynamics of this hybrid dynamical system, we consider the solution starting from a guard condition at the initial time $t = t_n$. Solving (3.11) admits an initial solution

$$u(t) = s_n + (u_n - s_n) e^{\beta(t - t_n)} \left\{ \cos \omega(t - t_n) - \frac{\beta}{w} \sin \omega(t - t_n) \right\}, \tag{3.18}$$

which is valid for $t \ge t_n$ as long as the discrete state $s(t)$ remains unchanged. The discrete state can only change when the guard condition $\frac{du}{dt}(t) = 0$ is satisfied. For the initial solution (3.18), the derivative is

$$\frac{du}{dt}(t) = -\frac{\omega^2 + \beta^2}{\omega} (u_n - s_n) e^{\beta(t - t_n)} \sin \omega(t - t_n), \tag{3.19}$$

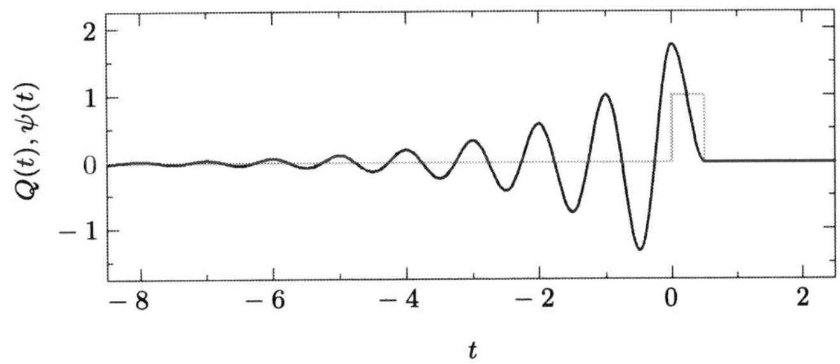

Figure 3.14. *Basis pulses $Q(t)$ (black) and $\psi(t)$ (gray) for the exact folded-band oscillator with $\beta = 0.81 \cdot \ln 2$. Figure courtesy of US Army.*

so that the guard condition is next met at $t = t_n + 1/2$ (recalling $\omega = 2\pi$). Evaluating the solution (3.18) at this time gives

$$u(t_n + 1/2) = s_n - e^{\beta/2}(u_n - s_n) \tag{3.20}$$

which is true for any initial condition corresponding to a guard condition. Since the system (3.11)–(3.12) is autonomous, the relationship (3.20) can be translated in time and used to propagate a solution from one guard condition to the next.

Using the relationship (3.20), we can then derive an analytic expression for a successive maxima return map in the continuous state waveform. To this end, we consider an initial condition above such that $u(t_n)$ is also a local maximum. This additional specification implies the trajectory initially decreases, so that $\frac{du}{dt} < 0$ for $0 < t - t_n < \frac{1}{2}$ in (3.19), which requires $u_n > s_n$. Consequently, any local maximum in the waveform is necessarily positive.

We first consider an initial local maximum such that $0 < u_n \leq 1$ and $s_n = 0$. Using (3.20), we find

$$u(t_n + 1/2) = -e^{\beta/2} u_n < 0 \tag{3.21}$$

so that, by (3.12), we also have $s(t_n + 1/2) = 0$. Applying (3.20) again yields

$$u(t_n + 1) = e^{\beta} u_n > 0 \tag{3.22}$$

which is the next local maximum. The linear relationship in (3.22) provides the first segment (labeled A) of the successive local maxima return map shown in Figure 3.15. Since $\beta > 0$, the slope of this linear segment is necessarily greater than one.

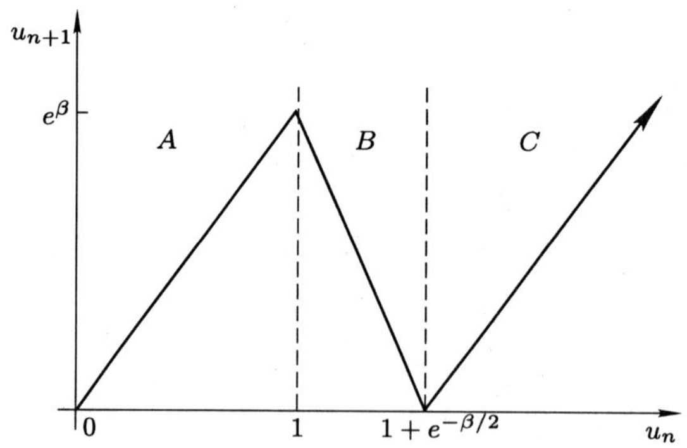

Figure 3.15. *Successive maxima return map for the exact folded-band oscillator. Figure courtesy of US Army.*

We next consider an initial maximum such that $u_n > 1$ and $s_n = 1$. Then (3.20) gives

$$u(t_n + 1/2) = 1 + e^{\beta/2}(1 - u_n) < 1 \tag{3.23}$$

so that $s(t_n + 1/2) = 0$. We again apply (3.20) to find

$$u(t_n + 1) = e^{\beta} u_n - (e^{\beta/2} + e^{\beta}), \tag{3.24}$$

where the sign of $u(t_n+1)$ depends on the magnitude of u_n. For $1 < u_n < 1 + e^{-\beta/2}$ we find that $u(t_n+1) < 0$ and $s(t_n+1) = 0$. In this case, applying (3.20) one more time yields

$$u(t_n + 3/2) = (e^{\beta} + e^{3\beta/2}) - e^{3\beta/2} u_n > 0, \tag{3.25}$$

which is the next local maximum. This relationship provides the second linear segment (labeled B) for the successive local maxima return map shown in Figure 3.15. The slope of this segment is negative, but its magnitude is greater than that of the first segment. For the case $u_n > 1 + e^{-\beta/2}$, we see that (3.24) gives $u(t_n+1) > 0$, which is the next local maximum. Thus (3.24) provides the third and final segment (labeled C) of the return map in Figure 3.15. The slope for this third segment is the same as the first.

The composite map relating successive maxima is then given by the one-dimensional, piecewise-linear map

$$u_{n+1} = f(u_n), \tag{3.26}$$

where

$$f(u_n) = \begin{cases} e^{\beta} u_n, & 0 < u_n \le 1, \\ -e^{3\beta/2} u_n + (e^{\beta} + e^{3\beta/2}), & 1 < u_n < 1 + e^{-\beta/2}, \\ e^{\beta} u_n - (e^{\beta/2} + e^{\beta}), & u_n > 1 + e^{-\beta/2}, \end{cases} \tag{3.27}$$

which is shown in Figure 3.15. We note the return times for the continuous system are not evenly spaced. In fact, the return times depend on the system state as

$$t_{n+1} = \begin{cases} t_n + 1, & 0 < u_n \le 1, \\ t_n + 3/2, & 1 < u_n < 1 + e^{-\beta/2}, \\ t_n + 1, & u_n > 1 + e^{-\beta/2}, \end{cases} \tag{3.28}$$

and correspond to the three segments of the piecewise-linear return map.

Since the magnitude of the slope for the one-dimensional return map (3.27) is everywhere greater than one, any bounded trajectory of the iterated map (3.26) is necessarily chaotic. By implication, corresponding dynamics of the hybrid system are also chaotic. Furthermore, the one-dimensional map is not invertible, and neither are the dynamics of the continuous-time oscillator. Consequently, the hybrid dynamical system defines a chaotic semiflow, where the future is determined by the current state but the past is not uniquely defined.

For the iterated map (3.26), a necessary and sufficient condition for persistent chaos is

$$f(e^{\beta}) \le e^{\beta} \tag{3.29}$$

which implies the return map (3.27) is closed on the interval $0 \le u_n \le e^{\beta}$. Condition (3.29) is true if and only if $\beta \le \beta_2$, where

$$e^{\beta_2} = 2 + e^{-\beta_2/2}. \tag{3.30}$$

Since the left side of (3.30) is a monotonically increasing function of β_2, while the right side is a monotonically decreasing function, we see there exists a unique solution to (3.30). Numerically, we find $\beta_2 \cong 1.39 \cdot \ln 2$. Thus the hybrid system exhibits bounded chaotic dynamics for $0 < \beta \le \beta_2$.

The return map (3.27) shows three linear segments. However, the chaotic dynamics of the iterated map are restricted to just the first two segments (labeled A and B) if

$$e^{\beta} \leq 1+e^{-\beta/2}, \tag{3.31}$$

implying the map is closed on the interval $0 \leq u_n \leq 1+e^{-\beta/2}$. It is straightforward to show that (3.31) is satisfied for all $\beta \leq \beta_1$, where

$$e^{\beta_1} = 1+e^{-\beta_1/2}. \tag{3.32}$$

Similar to (3.30), there is also a unique solution to (3.32). Furthermore, we recognize $\beta_1 \leq \beta_2$, and we numerically find $\beta_1 \cong 0.81 \cdot \ln 2$. Consequently, the hybrid system exhibits simply folded-band dynamics characterized by a skew tent map for $0 < \beta \leq \beta_1$. The hybrid system exhibits more complex dynamics for $\beta_1 < \beta \leq \beta_2$, characterized by the full piecewise-linear map of three segments.

For $\beta > \beta_2$, the return map is not closed on any finite interval, and we do not expect oscillations of the hybrid system to be bounded. In this case, any chaotic dynamics in the hybrid system are necessarily transient [36].

In using the inverse coding function (3.17), there are grammar restrictions on the allowed amplitude sequences that can be targeted. To understand the set of allowable sequences, we consider the symbolic dynamics of the successive maxima return map. For the one-dimensional return map, we use an alphabet of three symbols (A, B, C) with the generating partition

$$S_n = \begin{cases} A, & 0 \leq u_n < 1, \\ B, & 1 \leq u_n < 1+e^{-\beta/2}, \\ C, & 1+e^{-\beta/2} \leq u_n, \end{cases} \tag{3.33}$$

as shown in Figure 3.15. For the special cases $\beta = \beta_1$ and $\beta = \beta_2$, the iterated map is a full shift on two (A, B) or three (A, B, C) symbols, respectively. In these cases, the grammar is unrestricted, meaning that any sequence of the symbols is allowed by the dynamics. For other values of β, the map is not a full shift and the grammar is restricted. For $\beta < \beta_1$, the map is unimodal and the two-symbol grammar is restricted according to a maximal kneading sequence [11]. For $\beta_1 < \beta < \beta_2$, an analogous maximal sequence may be similarly used to define the three-symbol grammar restriction [12].

For an allowable symbol sequence $\{S_n\}$, we identify the corresponding amplitude sequence $\{\sigma_m\}$ by considering the intermediate steps in the derivation of the successive maxima return map (3.27). Referring to (3.21) and (3.22), we see that an iterate corresponding to symbol A admits $s(t) = 0$ for one full time unit. Similarly, (3.23) and (3.24) reveal that both symbols B and C admit $s(t) = 1$ for a half time unit followed by $s(t) = 0$ for another half time unit. Equation (3.25) implies symbol B requires $s(t) = 0$ for an additional half time unit. Thus, the substitutions

$$\begin{aligned} A &\rightarrow 00, \\ B &\rightarrow 100, \\ C &\rightarrow 10 \end{aligned} \tag{3.34}$$

provide a mapping from a symbol sequence to an amplitude sequence, where each amplitude occurs for a half time unit. Consequently, for any allowed future symbol sequence, the mapping (3.34) enables the calculation of an initial condition (or control point) via (3.17). This composition provides the inverse coding function for the exact folded-band oscillator.

An example of implementing the inverse coding function is shown in Figure 3.16. In this example, we consider the initial condition corresponding to the symbol sequence

$$\{S_n\} = BABABAABAABA\bar{A}, \tag{3.35}$$

where the overbar indicates the sequence terminates with the infinite repetition of the symbol A. Using the substitutions (3.34), the corresponding amplitude sequence is

$$\{\sigma_m\} = 100\,00\,100\,00\,100\,00\,00\,100\,00\,00\,1\bar{0}. \tag{3.36}$$

For $\beta = 0.81 \cdot \ln 2$, the exact analytic solution computed using (3.14) is shown in the top plot of Figure 3.16. We note that the magnitude of successive maxima in this solution relative to a threshold at one agree with the sequence (3.35). The results of numerically integrating the hybrid dynamical system (3.11)–(3.12) using an initial condition calculated via (3.17) are shown in the middle plot of Figure 3.16. We see that the numerical solution initially matches the exact analytic solution, generating a similar waveform with the correct symbol sequence for $t < 11$. The magnitude of the difference between the analytic and numerical solutions for the continuous state is shown in the bottom plot of Figure 3.16. In this plot, we see a growing difference in the solutions, so that eventually the numerical solution does not generate the correct symbols for $t \geq 11$. We attribute the

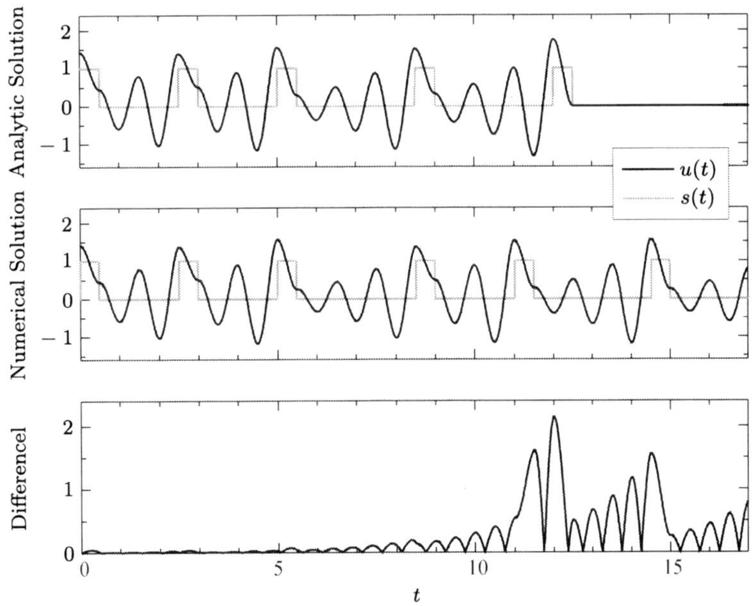

Figure 3.16. *Comparison of the exact analytic solution and a numerical integration of the exact folded-band oscillator with $\beta = 0.81 \cdot \ln 2$. Figure courtesy of US Army.*

divergence in these solutions to the limited precision in the numerical solution and the chaotic nature of the system.

Since the solution for this system is a linear convolution, the spectral properties of the chaotic continuous waveform can be inferred from the Fourier transform of the basis function. It is

$$q(\Omega)=\int_{-\infty}^{\infty} Q(t)e^{i\Omega t}dt=\frac{e^{i\Omega/2}-1}{i\Omega}\cdot\frac{\omega^2+\beta^2}{\omega^2+(\beta+i\Omega)^2}, \tag{3.37}$$

where $i=\sqrt{-1}$. We note this result is nearly the same as the Fourier transform $p(\Omega)$ calculated for the basis pulse in the exact shift oscillator. We address this relationship in the next subsection.

A circuit schematic showing an electronic folded-band oscillator is shown in Figure 3.17. Similar to the implementation of the shift oscillator, this circuit does not explicitly implement the guard condition but rather uses threshold comparators and simple logic circuits. Its simplicity again suggests that high-frequency implementations of the exact folded-band oscillator may also be easily realized.

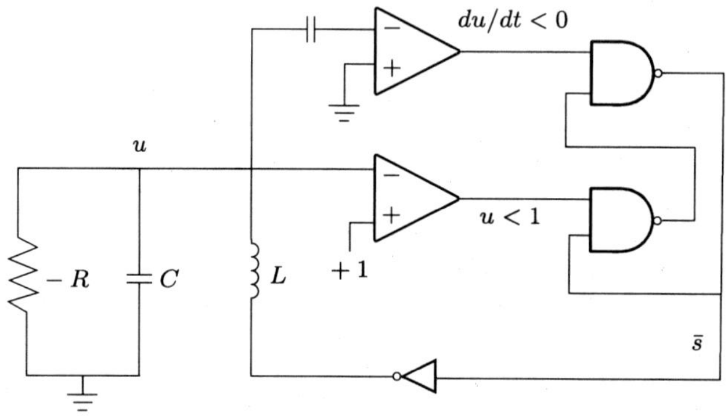

Figure 3.17. *Circuit realization of exact folded-band oscillator.*

3.3.3 ▪ Connection between Exact Oscillators

As already indicated, both exactly solvable oscillators can be derived from a single model of manifold piecewise-linear chaos [25]. As such, it is reasonable to expect an analytic connection may exist that relates the analytic solutions. In considering each solution, we recognize that

$$P(t)=Q(t)+Q\left(t-\frac{1}{2}\right), \tag{3.38}$$

where $P(t)$ and $Q(t)$ are the basis functions for the two oscillators and are given by (3.1)–(3.2) and (3.11)–(3.12), respectively. This connection follows since both hybrid systems incorporate the same differential equation to describe the continuous dynamics. In the frequency domain, the equivalent relationship is

$$p(\Omega)=(1+e^{i\Omega/2})\cdot q(\Omega), \tag{3.39}$$

where $p(\Omega)$ and $q(\Omega)$ are the Fourier transforms of the respective basis pulses. It is a straightforward (but not necessarily illuminating) calculation to verify these relationships. However, the necessity of the relationships is more obvious when the dynamics of both oscillators are considered in reverse time [31]. In previous work, it was shown that chaotic waveforms can be synthesized backward in time using a stable linear filter driven by a random square wave. While the Lorenz topology can be synthesized using random square waves of unit duration, synthesizing a folded band using the same filter requires random square waves with half-unit duration [32].

A consequence of the relationships (3.38) and (3.39) is that chaotic trajectories for the exact shift system can also be represented by the mathematical solution of the exact folded-band system. Thus, the only difference between the solutions for the two oscillators, which characterize two distinct topologies, is the set of allowable amplitude coefficients σ_m. In fact, each system admits a different allowable grammar in terms of the amplitude sequences σ_m, and it is the different grammar that distinguishes the topological structure of each oscillator. This grammar-dependent topology, and the existence of a chaotic superset incorporating both the Lorenz butterfly and Rössler's folded band, was discussed previously in the context of chaos synthesized in reverse time [32].

3.4 ▪ Symbolic Dynamics Control

In this section, we describe chaos communication by control of symbolic dynamics [13]. Our assumption is that we can use these methods with the exact systems described in the previous section. From an encoding point of view, the key advantage provided by the solvable oscillators is the availability of an exact symbolic dynamics, including an analytic representation of the inverse coding function required for chaos control. These oscillators also offer an advantage for the receiver, which we address in the following section.

3.4.1 ▪ Symbolic Dynamics

A chaotic attractor defines a set of trajectories that a chaotic oscillator may possibly follow. Normally, the choice of which particular trajectory the oscillator follows is arbitrary or random. However, an approach to chaos communications is to associate a message with each possible trajectory and then to coax the oscillator to follow the trajectory corresponding to a selected message to be transmitted. Although conceptually simple, this task of assigning trajectories to messages quickly explodes in its combinatorial options due to the exponential divergence inherent to chaotic dynamics. What transforms this potentially overwhelming task into an elegant approach is the use of a symbolic dynamics [9].

In the study of nonlinear dynamics and chaos, a symbolic dynamics is used to specify and label any continuous, real-valued trajectory of a chaotic oscillator using a sequence of discrete, finite-precision samples [11]. To define a symbolic dynamics, the state space containing the attractor is partitioned into regions, each of which is labeled with a discrete symbol. The choice of what to use as symbols is arbitrary, such as "0", "1", "L", "C", "R," etc., and the set of all discrete symbols is called an alphabet. Any particular trajectory then corresponds to a sequence of symbols indicating the order in which the regions are traversed. For flows represented by a return map, a convenient partition assigns a symbol to each return using a partition of the return surface. In this way any recurrent trajectory maps to an infinite sequence of symbols, with one symbol for each return.

Since a trajectory's symbol sequence specifies the order in which regions are visited, time is implied by the symbolic representation. We can then define a current time in the symbolic representation, which identifies the concurrent symbolic state of the system. That is, the current symbol indicates the region of state space in which the present state of the system resides. It is conventional to indicate the current symbol using a decimal point in the sequence, such as in the symbolic trajectory $\mathbf{a} = \ldots a_{-2}a_{-1}.\, a_0a_1a_2\ldots$, where a_0 is the current symbol, future symbols are written toward the right (a_i, $i > 0$), and past symbols to the left (a_i, $i < 0$). By construction, every trajectory on a chaotic attractor produces a unique symbol sequence. Likewise, every initial condition x corresponds to a trajectory, which can then be mapped to sequence space, including specification of the current symbol. In terms of chaos communication, this mapping is called the coding function, $\mathbf{a} = r(x)$ [9]. Effectively, the coding function assigns a symbol sequence to every point on the attractor.

Of course, the choice of the partition is critical to such a labeling, and a good partition will yield symbolic sequences that uniquely describe each trajectory using the smallest alphabet of symbols. State space and sequence space are topologically equivalent, provided the coding function is one-to-one and continuous. That is, every point on the attractor corresponds to a unique symbol sequence, and nearby points in state space are also nearby in sequence space. Although the concept of nearby points in state space is well understood, for this to make sense we need to quantify the notion of closeness in symbol space. For example, given two symbol sequences $\mathbf{a} = \ldots a_{-2}a_{-1}.\, a_0a_1a_2\ldots$ and $\mathbf{b} = \ldots b_{-2}b_{-1}.\, b_0b_1b_2\ldots$, a distance between them can be defined using

$$d(\mathbf{a},\mathbf{b}) = \frac{1}{2}\sum_{i=-\infty}^{\infty} \frac{\delta(a_i,b_i)}{2^{-|i|}}, \tag{3.40}$$

where

$$\delta(a_i,b_i) = \begin{cases} 0, & a_i = b_i, \\ 1, & a_i \neq b_i, \end{cases} \tag{3.41}$$

is the Kronecker delta function [37]. Thus sequences $\mathbf{a}$ and $\mathbf{b}$ are nearby if they agree in the first few symbols about the current symbol. A partition that guarantees the spaces are topologically equivalent is called generating, and finding a generating partition is straightforward for one-dimensional maps but is much more difficult in two or more dimensions [38, 39, 40]. In cases where an exact generating partition cannot be found, certain alternative partitions may still be effectively used [41, 42].

A key benefit of the symbolic representation is that it converts the complex dynamics of a chaotic oscillator into a trivial shift. To see this conversion, we consider an initial return at time $t = t_i$ such that the system state is $x(t_i)$. Using the coding function, we find the corresponding symbolic state is $\mathbf{a}_i = r(x(t_i))$. This symbolic representation indicates both the infinite sequence of partition symbols corresponding to the chaotic trajectory and the decimal point indicating the current partition symbol. The next return occurs at time $t = t_{i+1}$, to which the system state has evolved according to the oscillator dynamics to yield $x(t_{i+1})$. The corresponding symbolic state is $\mathbf{a}_{i+1} = r(x(t_{i+1}))$. However, since both $\mathbf{a}_i$ and $\mathbf{a}_{i+1}$ describe the same trajectory (i.e., they both came from the same initial condition), the symbol sequence in each representation must be the same. However, as time has progressed from t_i to t_{i+1}, the decimal point indicating current time must advance one symbol toward the future. Thus, the symbolic state $\mathbf{a}_{i+1}$ is computed directly from $\mathbf{a}_i$ by simply shifting each symbol one position to the left (equivalently, moving the decimal point one symbol to the right). Mathematically, we denote this operation by the shift operator,

$$\mathbf{a}_{i+1} = \sigma(\mathbf{a}_i), \tag{3.42}$$

which explicitly defines the dynamical system in the symbol-space representation.

Experimentally or numerically, one can observe only an approximate or truncated representation of the symbolic state $\mathbf{a}_i$ consistent with the measurement or numerical resolution. Topological equivalence means that specifying more future or past symbols defines the current state to a higher level of precision. For example, we might consider the finite symbolic description $\mathbf{a}_i$ ="10.1101110", which implies a resolution capable of distinguishing eight future and two past symbols. Now, as the system evolves to the next return, the shift dynamic (3.42) introduces new information that appears on the right-hand side of the truncated sequence [43]. Measured with the same resolution, the next return is $\mathbf{a}_{i+1}$ ="01.101110?", where "?" is a new symbol that cannot be predicted based on the limited knowledge of $\mathbf{a}_i$. One can think of this symbol as a previously undetectable detail of the initial condition that is now revealed by the chaotic dynamics. The continuation of the shifting process is illustrated in Figure 3.18, where the symbol sequences for subsequent returns are shown. In this process, a new symbol of information appears on the right with each return, and the initial uncertainty grows with time. Eventually, the uncertainty due to limited resolution completely replaces any measured knowledge of the system. At this time scale, prediction is rendered useless, which is a known hallmark of chaos.

3.4.2 ▪ Chaos Control

It is well known that ideal chaotic systems can be controlled using arbitrarily small perturbations [2, 7, 44]. The fundamental nature underlying this concept is easily and conveniently explained using symbolic dynamics. Generally, the job of chaos control

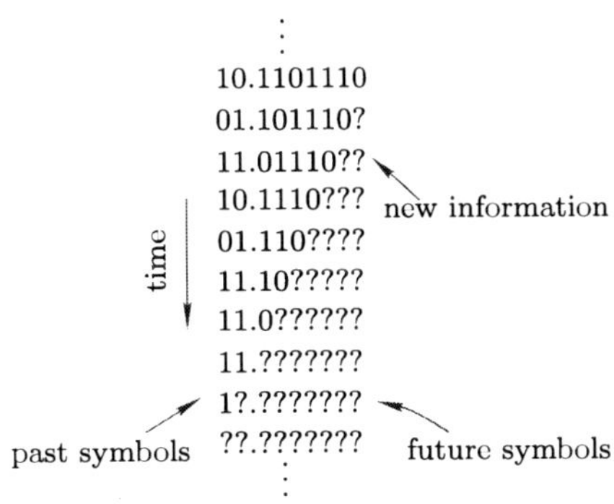

Figure 3.18. *Dynamical evolution of the symbolic state showing propagation of uncertainty due to limited precision.*

is to drive the system state to follow a particular, preselected trajectory. In terms of the symbolic representation, the controller replaces the uncertainty of future symbols with a particular sequence corresponding to the target trajectory. An example of symbolically targeting and controlling a period-two orbit is shown in Figure 3.19. By exploiting the shift dynamics, it is efficient to encode target symbols well into the future, where the requisite perturbations are small. As the system evolves, the symbolic shift operation naturally brings these controlled symbols to the macro scale. Using smaller control signals sets symbols further into the future, so the penalty is a longer time lag before the controlled symbols are exhibited by the system.

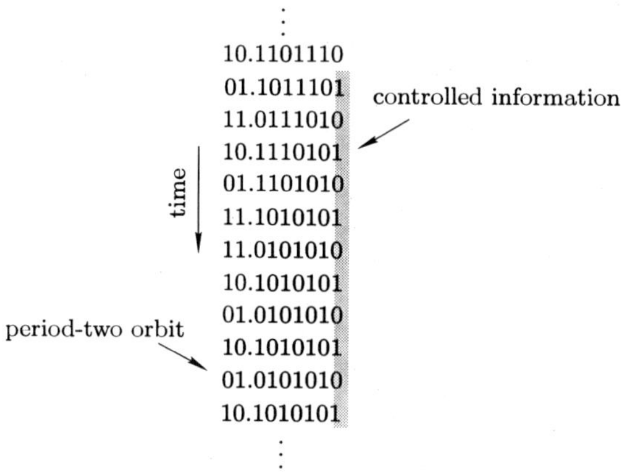

Figure 3.19. *Symbolic representation of chaos control targeting a period-two orbit.*

Historically, chaos control was originally proposed as a method to stabilize one of an infinite number of unstable periodic orbits (UPOs) in a chaotic attractor [2]. However, it was soon realized that aperiodic trajectories could also be stabilized using small perturbation control [9]. In terms of controlling symbolic dynamics, the difference is immaterial: repeating symbol sequences specify periodic orbits, but a nonrepeating symbol sequence specifies a chaotic trajectory [45, 46]. Indeed, any symbolic trajectory consistent with the chaotic dynamics (i.e., the allowable grammar) can be controlled.

In practice, targeting a symbolic trajectory requires specifying a sequence of control points in state space and applying a chaos controller to drive the trajectory through the control points [13]. The process of defining the control points requires converting each symbolic representation in the targeted symbolic trajectory to a corresponding state space value. If the partition is generating, the coding function is invertible and a control point is simply $x_i = r^{-1}(\mathbf{a}_i)$, where $\mathbf{a}_i$ is the ith return in the target symbolic state. In applying the inverse coding function to a truncated symbolic representation, the choice of the unknown future symbols beyond the control symbol is arbitrary and inconsequential, since the effect is smaller than the applied control perturbation.

In this picture, the exactly solvable oscillators shown in the previous section are exceptionally well suited to control of symbolic dynamics. Each oscillator admits an exact, one-dimensional return map, for which a symbolic dynamics with a generating partition is known. Thus, each oscillator has a convenient and well characterized symbolic dynamics. Furthermore, each oscillator exhibits an analytic inverse coding function, by which any future symbol sequence is analytically mapped to a control point. Finally, the grammar restrictions for each oscillator's symbolic dynamics are known in terms of the one-dimensional map dynamics.

3.4.3 ▪ Dynamic Limiting

In theory there exist a number of chaos control techniques that can be employed to steer the state of the oscillator to follow a sequence of control points. However, when considering an electronic implementation of a high-data rate communication system, it is a practical necessity to consider controllers best suited for fast control. A promising high-speed electronic chaos control technique is limiter control, where simple diode switches can be used to stabilize chaotic dynamics [47]. Previously it has been experimentally demonstrated that such control can be extended to dynamic limiting for stabilizing long period orbits and encoding symbolic dynamics [48].

Figure 3.20 shows an idealized circuit implementation of a static limiter applied to a generic chaotic LC oscillator. The limiter is implemented using a diode biased with a voltage source corresponding to the limiter level h. When the oscillator voltage u exceeds the limiter bias h, the diode turns on and excess energy is dissipated by the shunt resistor R_D. By setting h to the maximum peak return of an unstable periodic orbit, a static limiter can stabilize a few of the lowest order periodic orbits [47]. Stabilizing longer periodic orbits or arbitrary trajectories requires a dynamic limiter, where $h = h(t)$ contains a sequence of limiter levels that describe the successive maxima returns of the controlled orbit [48]. Figure 3.21 shows an idealized implementation of a dynamic limiter designed to control an arbitrary symbolic sequence. In this view, a sequence of future symbols is fed into a shift register, which is clocked by the return time of the oscillator and implements the shift dynamics of the symbolic state. The state stored in the shift register is converted to a control level via the inverse coding function. In this way, a sequence of control points for successive returns can be generated corresponding to a continuous symbol stream input to the controller. This method of symbolic dynamics control has been used previously for controlling folded-band oscillators in experimental systems to test the fundamental limits of chaos synchronization [49, 50].

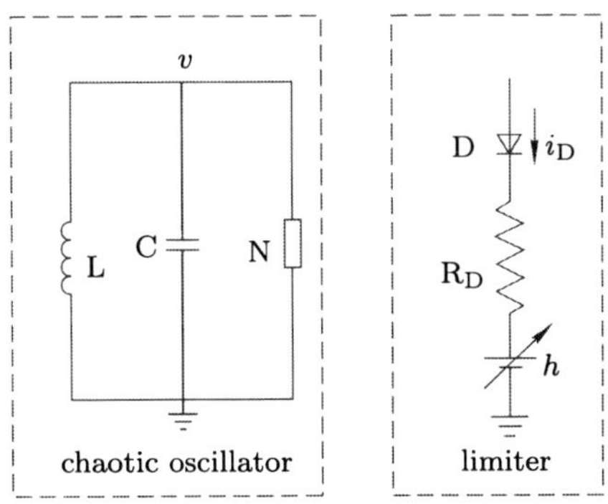

Figure 3.20. *Static limiter control of a chaotic LC electronic oscillator for stabilizing low-order unstable periodic orbits.*

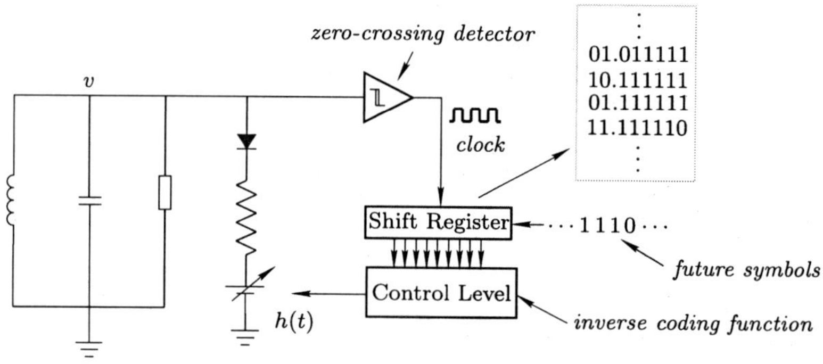

Figure 3.21. *Dynamic limiter control of a chaotic electronic oscillator. A sequence of future symbols describing a target trajectory is input to a shift register clocked by the oscillator return time and used to define a limiter control level for successive waveform maxima.*

To demonstrate control of symbolic dynamics, we simulate dynamic limiter control applied to the exact folded-band oscillator. To limit the oscillator state, we modify the continuous dynamics of the oscillator as

$$\frac{d^2u}{dt^2} - \{2\beta - G \cdot H(u-h)\} \cdot \frac{du}{dt} + (\omega^2 + \beta^2) \cdot (u - s) = 0, \tag{3.43}$$

where h is the limiter level, G is the limiter strength, and H is the Heaviside function defined in (3.13). For $u < h$, the limiter is inert and oscillator dynamics are consistent with the uncontrolled system. However, the limiter resists large oscillations of the system state by adding a significant positive damping to dissipate oscillation energy whenever $u(t) > h$. The dissipative effect increases with the limiter strength, for which $G = 50$ is a typical value to achieve control.

Figure 3.22 shows simulation results using dynamic limiting to control symbolic dynamics in the exact folded-band oscillator. In the simulation, dynamic limiting is used to encode the target symbol sequence AABABABBBBAAAABAABBAABABBBBABBABB, where A and B are the symbols previously defined for the exact folded-band oscillator. The target sequence is used to define a corresponding target amplitude sequence via the substitution (3.34). Conveniently, the symbolic dynamics and coding function for the exact folded-band oscillator are derived using a successive maxima return map; therefore, the control points given by the inverse coding function (3.17) directly provide the limiter levels $h(t)$ required by the dynamic limiter. The top plot in Figure 3.22 shows the applied dynamic limiter levels $h(t)$ and the resulting controlled continuous waveform $u(t)$. The controlled symbol sequence is shown above the plot, aligned to indicate the timing of each controlled symbol, and transitions in the dynamic limiter are triggered by negative-going zero crossings in the system state $u(t)$. The middle plot shows the corresponding switching state $s(t)$. The bottom plot shows the control perturbation strength, measured by the excess of the system state compared to the limiter level. Significantly, except for a transient when the oscillator state is initially captured, the control perturbation due to the limiter is small, implying the waveform is consistent with the dynamics of the uncontrolled oscillator.

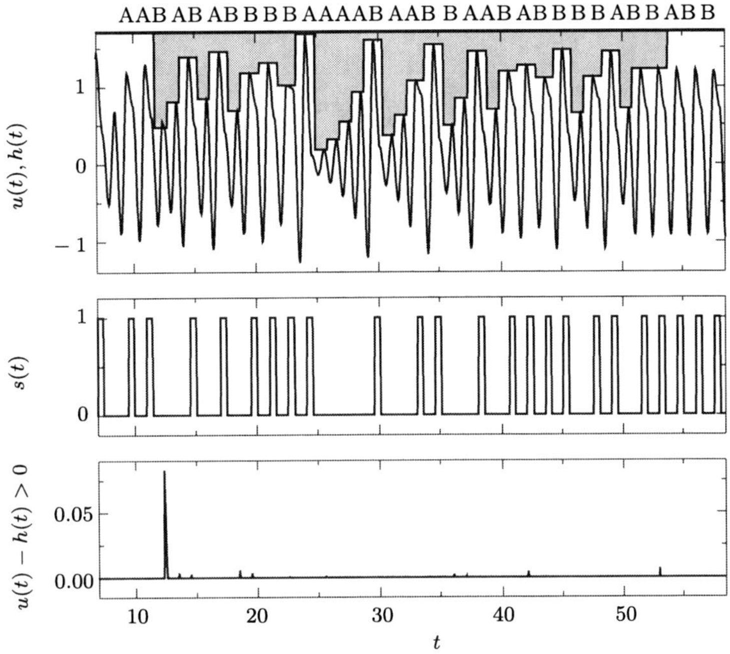

Figure 3.22. *Dynamic limiter control of the exact folded-band oscillator with* $\beta = 0.78 \cdot \ln 2$ *and using limiter strength* $G = 50$. *The top plot shows the continuous oscillator state* $u(t)$ *and dynamic limiter control levels* $h(t)$ *(gray). The middle plot shows the corresponding switching state* $s(t)$. *The bottom plot shows the relative control perturbations, which are small after an initial transient. The target symbol sequence is shown at top. Figure courtesy of US Army.*

3.4.4 ▪ Channel Capacity

In any communication system, a critical performance characteristic is the channel capacity, which indicates the maximum rate that information can be carried through the channel. In this subsection, we are concerned both with the capacity of a chaotic dynamical system to encode information and the requirements on a physical channel to transmit the encoded chaotic waveform. We begin by considering the channel capacity of the symbolic dynamics encoding.

Measured in bits of information per time, the capacity of a noiseless discrete channel is

$$C = \lim_{T\to\infty} \frac{\log_2 N(T)}{T}, \tag{3.44}$$

where $N(T)$ is the number of different allowed message signals of length T [22]. For coding symbolic dynamics, the capacity reflects the number of symbolically distinct trajectories that can be controlled using small perturbations. These trajectories must be consistent with the allowable grammar of the system. We note the lengths of the trajectories are measured in time, and the definition accounts for both the size of the alphabet and duration of individual symbols.

A fundamental theorem of information theory states that information can be transmitted through a channel if its entropy H is less than or equal to the channel capacity C.

In this context the entropy of a source is

$$H = \lim_{T\to\infty} \sum_{i=1}^{N(T)} -p_i \log_2(p_i), \tag{3.45}$$

where the sum is over all messages of length T and p_i is the probability of the ith message. An obvious example of a source compatible with the symbolic dynamics channel is the free running chaotic oscillator itself, using symbols derived with the same partition. In this case, the set of message probabilities reflects the natural invariant density of the uncontrolled oscillator. For a generating partition, the entropy derived from the symbolic dynamics is the metric entropy of the dynamical system, and chaos is characterized by positive metric entropy [51]. Loosely, the metric entropy quantifies the average amount of information revealed by successive measurements made of the evolving system using a consistent, limited resolution.

Comparing (3.44) and (3.45), we note that equality $H = C$ is achieved for the uniform probabilities $p_i = 1/N(T)$ for all i, and any deviation from uniform probabilities results in $H < C$. In other words, the oscillator uses the entire channel capacity if and only if all allowable messages are equally likely. This maximal entropy, which equally emphasizes all possible sequences, is called the topological entropy for a chaotic oscillator [51]. Thus, we see that the topological entropy of a chaotic oscillator is the channel capacity of a symbolic dynamics encoding using that oscillator [52]. In general, the topological entropy is an upper bound on the metric entropy, and any difference between them is due to tendencies that favor certain trajectories in an oscillator's natural invariant density. It may be curious to note that, by exploiting this difference, it is possible to encode information in a chaotic oscillator at a rate exceeding the entropy exhibited by the free-running system [53].

Although not rigorously proven in the general case, typical chaotic systems reasonably satisfy

$$H = \frac{1}{\ln 2} \sum_{\lambda_i > 0} \lambda_i, \tag{3.46}$$

meaning the metric entropy is equal to the sum of the positive Lyapunov exponents for the system [51]. (The multiplicative factor converts the natural-base exponents to bits). A simple heuristic argument explains this useful relationship. We note that positive Lyapunov exponents describe the average rate of expansion of any uncertainty in our knowledge of a chaotic system. That is, given an initial state measured with limited precision, our uncertainty in the system state grows exponentially with time per the positive Lyapunov exponents. A subsequent measurement taken with the same precision will collapse our uncertainty, effectively revealing new information about the system state. On average, the amount of information gained in each measurement is, therefore, simply the metric entropy as described above.

For both the exact shift and folded-band oscillators, we previously recognized the positive Lyapunov exponent is given by the negative damping β; thus, the calculation of the metric entropy is greatly simplified, yielding $H = \beta/\ln 2$. (Again, the denominator converts units to bits.) Since the topological entropy must be at least as large as the metric entropy, the negative damping also provides a lower bound for the channel capacity. That is, we must have $C \geq \beta/\ln 2$. However, we have strong reasons to expect the topological and metric entropies are the same for these oscillators. First, numerical evidence suggests the attractors for each system exhibit a uniform density. Second, each oscillator is characterized by a constant divergence due to the fixed negative damping. As such, we conjecture that the channel capacity for a symbolic dynamics encoding is $C = \beta/\ln 2$, although we have yet to definitively show this.

We now briefly consider the channel capacity required to transmit the encoded chaotic waveform through a physical channel. It has been shown previously that a channel capacity exceeding the metric entropy of an uncontrolled chaotic oscillator is necessary and sufficient to unambiguously specify a trajectory [54]. For a controlled oscillator, a channel capacity exceeding the topological entropy is sufficient and, for a maximal coding capacity, necessary. Theoretically, as long as the channel capacity is sufficient, the information can get through without errors. However, imperfections in the physical channel such as noise or interference can reduce the capacity of any communications system [22]. Practically, and especially in the case of insufficient channel capacity, the challenge remains of optimally recovering the signal and estimating the received message. We address this challenge in the next section by proposing a matched filter receiver for the exactly solvable chaotic oscillators.

As a final note regarding channel capacity, it has been previously suggested that, by restricting the grammar of the controlled system, the dynamics of the controlled oscillator can be kept from approaching states near the partition [13, 52]. In this way, a certain resistance to noise is achieved, since symbols near the partition are presumably difficult to discriminate and correspondingly more susceptible to noise. This approach, which arbitrarily limits the entropy used for encoding, achieves noise resistance at the cost of a decreased channel capacity. However, it is perhaps more efficient to use the entire channel capacity at the chaos level and rely on redundancy in established coding methods at the message-to-symbol level for error detection and correction [14].

3.5 ▪ Matched Filter Receiver

For communications using chaos, we see that small-perturbation control can be used to efficiently encode a message in the uncertain symbols that emerge with each return in the symbolic trajectory [9]. There is an undeniable elegance and efficiency in exploiting chaos for this purpose. However, the ability to exploit this method for practical communications is limited by the ability to effectively recover the information on reception. Surely, if the transmitted waveform is received unaltered, the symbolic information can be recovered by simply "looking" at the waveform returns and extracting symbols relative to the partition. However, practical reception is complicated by the presence of noise, interference, and other effects in a real communication channel. For effective communications, it is essential for the receiver to have a capability to minimize and overcome these detrimental effects.

In practice, optimal receivers generally rely on linear correlations to suppress noise and detect transmitted signals. Ideally, one would use a coherent receiver, which exploits an exact copy of the transmitted waveform used to encode each symbol [15]. The best example is a matched filter receiver, which provides optimal detection of the fixed basis signals in the presence of additive white Gaussian noise [55]. Normally, the uncertainties of typical chaotic waveforms would seem to deny a simple matched filter [15]. However, the linear superposition using a fixed basis function that underlies exactly solvable chaos is not typical and, in fact, enables a simple matched filter for these chaotic waveforms [20]. In this section, we describe this matched filter for chaos and indicate its use for detecting symbols in a chaos communications system.

3.5.1 ▪ Matched Filter

A matched filter for a given waveform is the optimal linear filter for detecting that waveform in AWGN [55]. Practically, a matched filter can be realized as a filter with an impulse response that is the time reversal of the waveform to be detected. The operation of a matched filter on a received waveform is mathematically equivalent to a correlation with the basis function.

For the exactly solvable shift oscillator, a matched filter for the basis function $P(t)$ is the linear operator Λ_P that provides

$$\mathscr{L}_P \circ \delta(t) = P(-t), \tag{3.47}$$

where $\delta(t)$ is the Dirac delta function. It has been shown that this operator is realized by the stable filter

$$\begin{aligned} \dot{\eta} &= v(t+1) - v(t), \\ \ddot{\xi} + 2\beta\dot{\xi} + (\omega^2+\beta^2)\xi &= (\omega^2+\beta^2)\eta(t), \end{aligned} \tag{3.48}$$

where $v(t)$ is the input, $\eta(t)$ is an intermediate state, and $\xi(t)$ is the output [20]. It is straightforward to verify that the response to an impulse excitation $v(t) = \delta(t)$ yields the exact solution $\xi(t) = P(-t)$. As such, (3.48) provides a matched filter for the basis function $P(t)$, and the matched filter can be used to detect waveform symbols in the presence of noise.

For the exact folded-band oscillator, the corresponding matched filter for the basis function $Q(t)$ is defined as

$$\mathscr{L}_Q \circ \delta(t) = Q(-t). \tag{3.49}$$

By exploiting the relationship between $P(t)$ and $Q(t)$, it is straightforward to find that the linear filter

$$\begin{aligned} \dot{\eta} &= v(t+1/2) - v(t), \\ \ddot{\xi} + 2\beta\dot{\xi} + (\omega^2+\beta^2)\xi &= (\omega^2+\beta^2)\eta(t) \end{aligned} \tag{3.50}$$

realizes the matched filter Λ_Q.

Figure 3.23 shows simulation results that illustrate the operation of the matched filter Λ_P operating on a received waveform in the presence of noise. The top plot shows the chaotic waveforms $u(t)$ and $s(t)$ generated by the exact shift oscillator with $\beta = \ln 2$. In this case, the oscillator is uncontrolled and free running; however, the waveforms could equally have been generated by a controlled oscillator encoding the observed symbolic sequence. The middle plot shows a received waveform $v(t)$ sampled with ten samples per time unit, where each sample includes additive Gaussian noise with mean $\mu = 0$ and standard deviation $\sigma = 0.5$. Here, a sampled waveform is used to limit the bandwidth and power of the AWGN process. The bottom plot shows the result of processing the noisy sampled waveform using the matched filter (3.48), for which a continuous-time input is

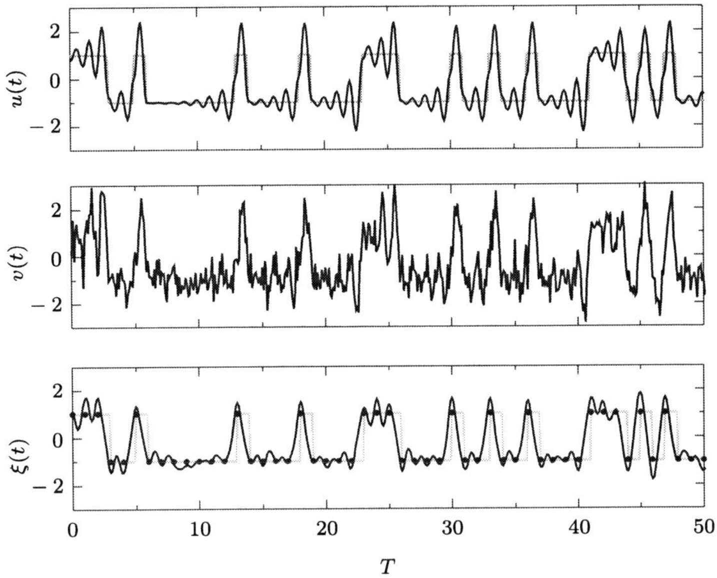

Figure 3.23. *Simulation results demonstrating a matched filter for the exact shift oscillator with $\beta = \ln 2$. The top plot shows the transmitted waveform $u(t)$ and the corresponding switching state $s(t)$. The middle plot shows the noisy sampled received waveform $v(t)$ with 10 samples per unit time and additive Gaussian noise ($\mu = 0, \sigma = 0.5$). The bottom plot shows the output of the matched filter $\xi(t)$, with return sample times indicated by dots and the transmitter switching state $s(t)$ in gray for reference.*

reconstructed using linear interpolation. We note that the function of the filter is not to recover the original transmitted waveform; instead, the matched filter is optimized to detect the symbol content of the transmitted waveform. The simulation results show the effectiveness of the filter for detecting the transmitted symbols, represented by the transmitter state $s(t)$ shown in gray in both the top and bottom plots. In the bottom plot we see that, by using a threshold at zero, the output of the matched filter sampled at returns times (dots) correctly identify the symbol content despite the presence of significant noise in the received waveform.

3.5.2 ▪ Performance

To analyze the effectiveness of a matched filter receiver, we consider the action of the matched filter Λ_p on a received waveform

$$v(t) = u(t) + w(t) \tag{3.51}$$

where $u(t)$ is a transmitted waveform from the exact shift oscillator and $w(t)$ is white Gaussian noise. Using properties of the delta function, we find that

$$\begin{aligned}\xi(t) = &\sum_{m=-\infty}^{\infty} s_m \int_{-\infty}^{\infty} P(\tau)P(\tau - t + m)d\tau \\ &+ \int_{-\infty}^{\infty} w(\tau)P(\tau - t)d\tau\end{aligned} \tag{3.52}$$

is the continuous-time output of the matched filter.

To detect a symbol s_n using the matched filter, the output $\xi(t)$ is sampled at $t = t_n$ and compared to a threshold. Here, we assume the timing problem has been solved and the output can be sampled at the precise time. As such, the sampled filter output is

$$\xi(t_n) = s_n I_0 + \Sigma_n + N, \tag{3.53}$$

where

$$I_0 = 1 + (1 - e^{-\beta}) \frac{\omega^2 - 3\beta^2}{2\beta(\omega^2 + \beta^2)} \tag{3.54}$$

is proportional to the energy contained in the basis function;

$$\Sigma_n = (e^{\beta} - 1) \frac{1 - I_0}{2} \sum_{\substack{m=-\infty \\ m \neq 0}}^{\infty} s_{n+m} e^{-|m|\beta} \tag{3.55}$$

is recognized as intersymbol interference; and N is the filtered noise. An estimate for the detected bit is then

$$\tilde{s}_n = \text{sgn}(\xi(t_n) - \theta), \tag{3.56}$$

where θ is the threshold. The probability of detecting an incorrect bit is

$$\begin{aligned} P_{BE} &= \Pr(\xi(t_n) < \theta | s_n = 1) \cdot \Pr(s_n = 1) \\ &\quad + \Pr(\xi(t_n) \geq \theta | s_n = -1) \cdot \Pr(s_n = -1), \end{aligned} \tag{3.57}$$

which is also known as the bit error rate (BER). In general, the threshold θ can be adjusted to minimize the BER. In particular, an optimal threshold will exploit a priori information about the current symbol s_n, as well as any knowledge of past and future symbols in Σ_n.

As a simple first case, we assume no prior knowledge of the current symbol, but all other symbols are known with certainty. That is, we assume

$$\Pr(s_n = 1) = \Pr(s_n = -1) = 1/2, \tag{3.58}$$

and the intersymbol interference Σ_n is fixed and known. In this case, we find

$$P_{BE} = \frac{1}{2} \{ \Pr(N < \theta - \Sigma_n - I_0) + \Pr(N > \theta - \Sigma_n + I_0) \}. \tag{3.59}$$

Assuming the noise N is normally distributed with zero mean and standard deviation σ, we find that optimal detection yields

$$P_{BE} = \frac{1}{2} \text{erfc} \left(\frac{I_0}{\sigma \sqrt{2}} \right) \tag{3.60}$$

using the threshold $\theta = \Sigma_n$. We note that (3.60) is equivalent to the standard result for BPSK, where

$$\frac{E_b}{N_0} = \frac{I_0^2}{2\sigma^2} \tag{3.61}$$

is the effective ratio of bit energy to noise power spectral density [14]. That is, if we can subtract the known intersymbol interference, detecting symbols is effectively equivalent to BPSK. In a sense, this knowledge provides a best case scenario, although detection can be further improved if a priori information about s_n is available.

In general, we must allow for uncertainty in the intersymbol interference Σ_n. In this case, we have

$$\begin{aligned} P_{BE} &= \Pr(s_n = 1) \cdot \int \Pr(\xi(t_n) < \theta | s_n = 1, \Sigma_n) \rho_+(\Sigma_n) d\Sigma_n \\ &\quad + \Pr(s_n = -1) \cdot \int \Pr(\xi(t_n) \geq \theta | s_n = -1, \Sigma_n) \rho_-(\Sigma_n) d\Sigma_n \end{aligned} \tag{3.62}$$

where $\rho_\pm$ designates a density function for Σ_n given $s_n = \pm 1$, respectively. In (3.62), the integrals are evaluated over the range of all possible values for the intersymbol interference in (3.55). As such, both density functions $\rho_\pm$ reflect the relative likelihood of the past and future symbol sequences $\{s_m\}$. Assuming the expected sequence is obtained from a free-running oscillator, this information encompasses the symbolic grammar of the oscillator, including disallowed sequences, as well as the relative frequency derived from the oscillator's invariant measure. For the general case, this information cannot be analytically derived, but rather it must be numerically estimated. If the sequence is generated by a controlled oscillator, as it would be in a communication system, the densities not only are properties of the oscillator but encompass information about the encoded message as well.

However, the special case of a free-running oscillator with $\beta = \ln 2$ does allow analytic treatment [20]. In this case, the return map (3.8) is a full shift, implying that the grammar is unrestricted and the symbols are independent and equally likely. As a result, the intersymbol interference is uniformly distributed over the interval $|\Sigma_n| \leq I_0 - 1$ and the integrals in (3.62) can be evaluated. For the optimal threshold $\theta = 0$, this calculation yields

$$\begin{aligned} P_{BE} = \frac{1}{4(I_0-1)^2} &\left\{ [(2I_0-1)^2 + \sigma^2] \operatorname{erfc}\left(\frac{2I_0-1}{\sigma\sqrt{2}}\right) - 2(I_0^2+\sigma^2) \right. \\ &\left. \operatorname{erfc}\left(\frac{I_0}{\sigma\sqrt{2}}\right) + (1+\sigma^2)\operatorname{erfc}\left(\frac{1}{\sigma\sqrt{2}}\right) \right\} \\ &- \frac{\sigma}{4(I_0-1)^2}\sqrt{\frac{2}{\pi}} \left\{ (2I_0-1)\exp\left(-\frac{(2I_0-1)^2}{2\sigma^2}\right) \right. \\ &\left. - 2I_0 \exp\left(-\frac{I_0^2}{2\sigma^2}\right) + \exp\left(-\frac{1}{2\sigma^2}\right) \right\} \end{aligned} \tag{3.63}$$

which provides an exact analytic expression for the BER in this case. In Figure 3.24, we compare this result with BPSK for $\beta = \ln 2$. In the figure, the horizontal axis is the effective ratio of bit energy to noise power spectral density using (3.61). In a sense, this result represents a worst case scenario for the matched filter receiver, since the intersymbol interference is the least constrained. However, the performance shown in Figure 3.24 still compares favorably to other correlation-based chaos communications schemes [15].

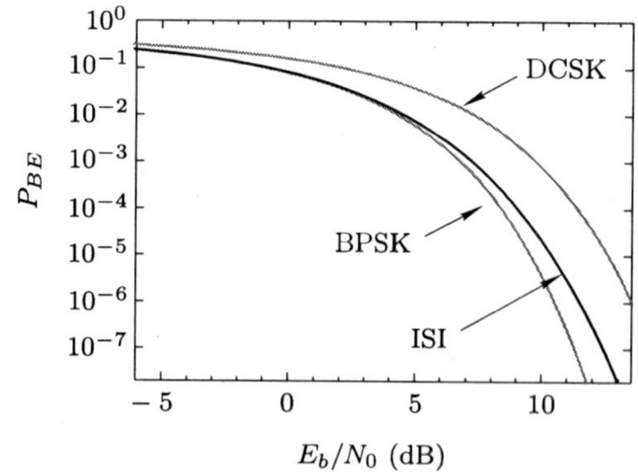

Figure 3.24. *Analytic BER for matched filter detection and* $\beta = \ln 2$ *with complete (BPSK) and no knowledge (ISI) about intersymbol interference. BPSK also represents the theoretical optimal limit. The best-case BER for coherent differential chaos shift keying (DCSK) is shown for comparison* [15].

3.6 ▪ Conclusions

The developments presented in this chapter are correctly viewed as a work in progress. Gaps remain in the description and characterization of this approach to communication with exactly solvable chaos. An objective of future work is to implement a prototype communication system, from which a proof-of-principle demonstration and performance metrics can be shown. So far, we have assembled many of the pieces, and it appears reasonable that the pieces can fit together. Yet, as of this writing, it is still a goal of future research to realize the entire system.

The key pieces we have identified are the exactly solvable chaotic oscillators. These systems offer a convenient, exact symbolic dynamics that can be utilized in the Hayes approach for communicating with chaos. Analysis yields an exact solution in the form of a linear convolution of a fixed basis pulse and a sequence of binary symbols forming the symbolic dynamics. Chaos control of the symbolic dynamics is facilitated by analytically available coding functions, which enable simple calculation of control points for target sequences. Importantly, the existence of a fixed basis function underlying the chaotic waveforms enables realization of a simple matched filter. As such, communicating with exactly solvable chaos can theoretically achieve performance comparable to conventional coherent communications technologies.

Besides its importance for enabling chaos communications, the development of a simple matched filter for chaotic waveforms enables the development of chaos radar [56]. For this application, a matched filter for an extended symbolic sequence is readily derived, enabling coherent pulse compression using just the symbolic representation of the transmitted waveform. Compared to random signal radars that sample and store the actual transmitted waveform, chaos radar using this matched filter can achieve similar performance with significantly less waveform sampling and storage requirements.

Finally, we conclude by commenting on the suitability of chaos communications for practical applications. It is often assumed that a communication channel is a wireless, through-the-air, radio-frequency link; however, such channels are typically characterized by significant noise, interference, and multipath and spectral constraints. As such, an over-the-air broadcast channel may be inappropriate for an ultrawide bandwidth chaotic waveform. But there are other scenarios for which chaos may be better suited. Possible examples include wired connections, optical fibers, laser systems, and FM radio channels. Besides a technical demonstration of this approach, the identification of such an application is also an important objective for future effort.

References

[1] L. M. Pecora and T. L. Carroll, "Synchronization in chaotic systems," *Phys. Rev. Lett.*, 64, pp. 821–824, 1990.

[2] E. Ott, C. Grebogi, and J. A. Yorke, "Controlling chaos," *Phys. Rev. Lett.*, 64, pp. 1196–1199, 1990.

[3] H. Fujisaka and T. Yamada, "Stability theory of synchronized motion in coupled-oscillator systems," *Prog. Theoret. Phys.*, 69, pp. 32–47, 1983.

[4] V. S. Afraimovich, N. N. Verichev, and M. I. Rabinovich, "Stochastic synchronization of oscillation in dissipative systems," *Radiophys. Quantum Electron.*, 29, pp. 795–803, 1986.

[5] M. P. Kennedy, R. Rovatti, and G. Setti (eds.), *Chaotic Electronics in Telecommunications*, CRC Press, Boca Raton, 2000.

[6] L. E. Larson, L. S. Tsimring, and J.-M. Liu, *Digital Communications Using Chaos and Nonlinear Dynamics*, Springer-Verlag, Berlin, 2006.

[7] E. Schöll and H. G. Schuster (eds.), *Handbook of Chaos Control*, 2nd ed., Wiley-VCH, Weinheim, 2008.

[8] L. Kocarev and S. Lian (eds.), *Chaos-Based Cryptography*, Springer-Verlag, Berlin, 2011.

[9] S. Hayes, C. Grebogi, and E. Ott, "Communicating with chaos," *Phys. Rev. Lett.*, 70, pp. 3031–3034, 1993.

[10] S. Hayes, C. Grebogi, E. Ott, and A. Mark, "Experimental control of chaos for communication," *Phys. Rev. Lett.*, 73, pp. 1781–1784, 1994.

[11] P. Collet and J. P. Eckmann, *Iterated Maps of the Interval as Dynamical Systems*, Birkhauser, Boston, 1980.

[12] B.-L. Hao and W.-M. Zheng, *Applied Symbolic Dynamics and Chaos*, World Scientific, Singapore, 1998.

[13] E. M. Bollt, "Review of chaos communication by feedback control of symbolic dynamics," *Internat. J. Bifur. Chaos Appl. Sci. Engrg.*, 13, pp. 269–285, 2003.

[14] M. B. Pursley, *Introduction to Digital Communications*, Pearson Prentice Hall, Englewood Cliffs, NJ, 2005.

[15] G. Kolumbán, M. P. Kennedy, and L. O. Chua, "The role of synchronization in digital communications using chaos—Part II: Chaotic modulation and chaotic synchronization," *IEEE Trans. Circuits Syst. I*, 45, pp. 1129–1140, 1998.

[16] R. Brown, N. F. Rulkov, and N. B. Tufillaro, "Synchronization of chaotic systems: The effects of additive noise and drift in the dynamics of the driving," *Phys. Rev. E*, 50, pp. 4488–4508, 1994.

[17] C.-C. Chen and K. Yao, "Stochastic-calculus-based numerical evaluation and performance analysis of chaotic communication systems," *IEEE Trans. Circuits Syst. I*, 47, pp. 1663–1672, 2000.

[18] E. Rosa, Jr., S. Hayes, and C. Grebogi, "Noise filtering in communication with chaos," *Phys. Rev. Lett.*, 78, pp. 1247–1250, 1997.

[19] N. Sharma and E. Ott, "Synchronization-based noise reduction method for communication with chaotic systems," *Phys. Rev. E*, 58, pp. 8005–8008, 1998.

[20] N. J. Corron, J. N. Blakely, and M. T. Stahl, "A matched filter for chaos," *Chaos*, 20, 023123, 2010.

[21] N. J. Corron and J. N. Blakely, "Exact folded-band chaotic oscillator," *Chaos*, 22, 023113, 2012.

[22] C. E. Shannon and W. Weaver, *The Mathematical Theory of Communication*, University of Illinois Press, Urbana, IL, 1949.

[23] E. N. Lorenz, "Deterministic nonperiodic flow," *Journal of Atmospheric Science*, 20, pp. 130–141, 1963.

[24] O. E. Rössler, "An equation for continuous chaos," *Phys. Lett. A*, 57, pp. 397–398, 1976.

[25] T. Saito and H. Fujita, "Chaos in a manifold piecewise linear system," *Electron. Commun. Jpn.* 1, 64(10), pp. 9–17, 1981.

[26] S. Katsura and W. Fukuda, "Exactly solvable models showing chaotic behavior," *Phys. A*, 130, pp. 597–605, 1985.

[27] K. Umeno, "Method of constructing exactly solvable chaos," *Phys. Rev. E*, 55, pp. 5280–5284, 1997.

[28] D. F. Drake and D. B. Williams, "Linear, random representations of chaos," *IEEE Trans. Signal Process.*, 55, pp. 1379–1389, 2007.

[29] S. T. Hayes, "Chaos from linear systems: Implications for communicating with chaos, and the nature of determinism and randomness," *J. Phys. Conf. Series*, 23, pp. 215–237, 2005.

[30] Y. Hirata and K. Judd, "Constructing dynamical systems with specified symbolic dynamics," *Chaos*, 15, 033102, 2005.

[31] N. J. Corron, S. T. Hayes, S. D. Pethel, and J. N. Blakely, "Chaos without nonlinear dynamics," *Phys. Rev. Lett.*, 97, 024101, 2006.

[32] N. J. Corron, S. T. Hayes, S. D. Pethel, and J. N. Blakely, "Synthesizing folded band chaos," *Phys. Rev. E*, 75, 045201, 2007.

[33] N. J. Corron, "An exactly solvable chaotic differential equation," *Dyn. Contin. Discrete Impuls. Syst. Ser. A*, 16, pp. 777–788, 2009.

[34] T. Tsubone and T. Saito, "Stabilizing and destabilizing control for a piecewise-linear circuit," *IEEE Trans. Circuits Syst. I*, 45, pp. 172–177, 1998.

[35] P. Glendinning and N. Sidorov, "Unique representations of real numbers in non-integer bases," *Math. Res. Lett.*, 8, pp. 535–543, 2001.

[36] C. Grebogi, E. Ott, and J. A. Yorke, "Fractal basin boundaries, long-lived chaotic transients, and unstable-unstable pair bifurcation," *Phys. Rev. Lett.*, 50, pp. 935–938, 1983.

[37] R. L. Devaney, *Introduction to Chaotic Dynamical Systems*, Addison-Wesley, New York, 1989.

[38] R. L. Davidchack, Y.-C. Lai, E. M. Bollt, and M. Dhamala, "Estimating generating partitions of chaotic systems by unstable periodic orbits," *Phys. Rev. E*, 61, pp. 1353–1356, 2000.

[39] Y. Hirata, K. Judd, and D. Kilminster, "Estimating a generating partition from observed time series: Symbolic shadowing," *Phys. Rev. E*, 70, 016215, 2004.

[40] M. Buhl and M. B. Kennel, "Statistically relaxing to generating partitions for observed time-series data," *Phys. Rev. E*, 71, 046213, 2005.

[41] E. M. Bollt, T. Stanford, Y.-C. Lai, and K. Zyczkowski, "What symbolic dynamics do we get with a misplaced partition? On the validity of threshold crossings analysis of chaotic time-series," *Phys. D*, 154, pp. 259–286, 2001.

[42] M. B. Kennel and M. Buhl, "Estimating good discrete partitions from observed data: Symbolic false nearest neighbors," *Phys. Rev. Lett.*, 91, 084102, 2003.

[43] R. Shaw, "Strange attractors, chaotic behavior, and information flow," *Z. Naturforsch.*, 36A, pp. 80–112, 1981.

[44] S. Boccaletti, C. Grebogi, Y.-C. Lai, H. Mancini, and D. Maza, "The control of chaos: theory and applications," *Phys. Rep.*, 329, pp. 103–197, 2000.

[45] E. M. Bollt and E. J. Kostelich, "Optimal targeting of chaos," *Phys. Lett. A*, 245, pp. 399–406, 1998.

[46] N. J. Corron and S. D. Pethel, "Experimental targeting of chaos via controlled symbolic dynamics," *Phys. Lett. A*, 313, pp. 192–197, 2003.

[47] N. J. Corron, S. D. Pethel, and B. A. Hopper, "Controlling chaos with simple limiters," *Phys. Rev. Lett.*, 84, pp. 3835–3838, 2000.

[48] N. J. Corron and S. D. Pethel, "Control of long-period orbits and arbitrary chaotic trajectories using dynamic limiting," *Chaos*, 12, pp. 1–7, 2002.

[49] N. J. Corron, S. D. Pethel, and K. Myneni, "Synchronizing the information content of a chaotic map and flow via symbolic dynamics," *Phys. Rev. E*, 66, 036204, 2002.

[50] S. D. Pethel, N. J. Corron, Q. R. Underwood, and K. Myneni, "Information flow in chaos synchronization: Fundamental tradeoffs in precision, delay, and anticipation," *Phys. Rev. Lett.*, 90, 254101, 2003.

[51] E. Ott, *Chaos in Dynamical Systems*, Cambridge University Press, Cambridge, 1993.

[52] E. Bollt, Y.-C. Lai, and C. Grebogi, "Coding, channel capacity, and noise resistance in communicating with chaos," *Phys. Rev. Lett.*, 79, pp. 3787–3790, 1997.

[53] M. B. Kennel and S. D. Pethel, "Complete and flexible replacement of chaotic uncertainty with transmitted information," *Phys. Rev. E*, 72, 046212, 2005.

[54] T. Stojanovski, L. Kocarev, and R. Harris, "Applications of symbolic dynamics in chaos synchronization ," *IEEE Trans. Circuits Syst. I*, 44, pp. 1014–1018, 1997.

[55] G. L. Turin, "An introduction to matched filters," *IRE Trans. Inform. Theor.*, 6, pp. 311–329, 1960.

[56] J. N. Blakely, N. J. Corron, and M. T. Stahl, "Concept for low cost chaos radar using coherent reception," *Proc. SPIE*, 8021, 80211H, 2011.

Chapter 4

Logic from Dynamics

William L. Ditto, Abraham Miliotis, K. Murali, and Sudeshna Sinha

Abstract: Here we review our work, spanning more than a decade, in developing the paradigm of *chaos computing*. The central idea is to exploit the rich temporal patterns embedded in the time evolution of nonlinear systems to obtain all logic functions in a flexible manner. This enables a single nonlinear element to be reconfigured into a variety of logic gates. So arrays of such morphing chaotic logic gates can then be programmed to perform higher-order functions, and to rapidly switch between such functions. Thus this reconfigurable chip will allow us to achieve the flexibility of field programmable gate arrays, the optimization and speed of application specific integrated circuits, and the general utility of a central processing unit within the same computer chip architecture. Further we present new results on the direct implementation of a set-reset flip-flop. Last, we suggest the direct emulation of basic logic operations by a one-dimensional and two-dimensional cellular automata. In this architecture, the cellular automata have the property that an initial excitation at a cell position depends upon the logical input and evolves quickly to yield different logic gate outputs. We call this cellular automata *logical cellular automata*, and it is further tested to do simple bit-by-bit arithmetic operations.

4.1 ▪ Introduction

Computation deals with finding a solution to a problem from given inputs by means of an algorithm, with tools ranging from pen and paper to electronic devices. Modern-day computers largely implement binary digital computing which is based on Boolean algebra, which shows how you can calculate anything with a system of two discrete values. For instance, it can be rigorously shown that any logic gate can be obtained by adequate connection of NOR or NAND gates (i.e., any Boolean circuit can be built using

NOR/NAND gates alone). This implies that the capacity for universal computing can simply be demonstrated by the implementation of the fundamental NOR or NAND gates [1].

The operation of any computing machine is necessarily a physical process, and this crucially determines the possibilities and limitations of the computing device. However, as conventional chip manufacturing technology runs into physical limits in the density of circuitry and signal speed, which sets limits to binary logic switch scaling, alternatives to semiconductor based binary digital computers are emerging, such as biochips, optical computers, and quantum computers.

The unifying idea in these alternate computing paradigms is the utilization of some of the computational capabilities inherent in the basic laws of physics. Thus these attempts aim to create machines that benefit from the basic laws of physics and are not just constrained by them. Here we review another emerging computing paradigm: one which exploits the richness and complexity inherent in nonlinear dynamics. This endeavor seeks to extend the possibilities of computing machines by utilizing the nonlinear features of the dynamics of the physical device.

It was proposed in 1998 that chaotic systems may be utilized to design computing devices [2]. In the early years the focus was on proof-of-principle schemes that demonstrated the capability of chaotic elements to do universal computing. The distinctive feature of this alternate computing paradigm was that they exploited the temporal patterns of chaotic systems.

In subsequent years, there has been much research activity to develop this paradigm [3–18]. It was realized that one of the most promising directions of this computing paradigm is its ability to exploit a single chaotic element to reconfigure into different logic gates through a threshold based morphing mechanism [3, 4]. In contrast to a conventional field programmable gate array element [19], where reconfiguration is achieved through switching between multiple single purpose gates, reconfigurable chaotic logic gates (RCLG) are composed of nonlinear elements that morph (or reconfigure) logic gates through the control of the pattern inherent in their dynamics. RCLGs have recently been realized and shown to be capable of reconfiguring between all logic gates in discrete circuits [5, 6, 7]. Additionally such RCLGs have been realized in prototype VLSI circuits (0.13μm CMOS, 30 MHz clock cycles). Further, reconfigurable chaotic logic gates arrays (RCGA) which morph between higher order functions such as those found in a typical arithmetic logic unit, have also been designed [17]. Thus we hope that our reconfigurable chaotic computer chips will enable us to achieve the flexibility of field programmable gate arrays, the optimization and speed of application specific integrated circuits (ASIC), and the general utility of a central processing unit (CPU) within the same computer chip architecture.

4.2 ▪ Review of the Chaos Computing Paradigm

First we recall the theoretical concept underlying the reconfigurable implementation of all fundamental logical operations utilizing both discrete-time and continuous-time nonlinear dynamics [3]. Then we will present recent results on the implementation of the fundamental set-reset flip-flop using a single chaotic Chua's circuit. We will show how our approach directly obtains outputs from the system by simple threshold mechanisms and does not necessitate the cross-coupling of NOR or NAND logic gates. Last, we will review a method to use synchronization to implement logic gates.

4.2.1 ▪ Logic from Threshold Controlled Chaotic Dynamics

We first recall the theoretical method for obtaining all basic logic gates with a single nonlinear system using threshold control. The broad aim here is to use the rich temporal patterns embedded in a nonlinear time series in a controlled manner to obtain a computing device that is flexible and reconfigurable.

Consider a chaotic element (our *chaotic chip* or *chaotic processor*) whose state is represented by a value x. In our scheme all the basic logic gate operations (NAND, NOR, XOR, AND, OR, XNOR, and NOT) involve the following steps:

(1) Inputs:

$x \to x_0 + X_1 + X_2$ for 2-input logic operations, such as the NAND, NOR, XOR, AND, OR, and XNOR operations,

and

$x \to x_0 + X$ for 1-input operations, such as the NOT operation.

Here x_0 is the initial state of the system, and

$X = 0$ when logic input is 0

and

$X = I_{in}$ when logic inut is 1,

where I_{in} is the input strength.

(2) Dynamical update, i.e., $x \to f(x)$,

where $f(x)$ is a nonlinear function.

(3) Threshold mechanism to obtain output Z:

$Z = 0$ if $f(x) \leq x^*$, and

$Z = f(x) - x^*$ if $f(x) > x^*$,

where x^* is a monitoring threshold.

This is interpreted as logic output 0 if $Z = 0$ and logic output 1 if $Z > 0$ (with $Z \sim I_{in}$).

Since the system is strongly nonlinear, in order to specify the initial x_0 accurately one needs a controlling mechanism. Here we will employ a threshold controller [20, 21] to set the initial x_0. Namely, we will use the clipping action of the threshold controller to achieve the initialization and subsequently to obtain the output as well.

Note that in our implementation we demand that the *input and output have equivalent definitions* (i.e., 1 unit is the same quantity for input and output), as well as among various logical operations. This requires that constant x^* assumes the same value throughout a network, and this will allow the output of one gate element to easily couple to another gate element as input, so that gates can be wired directly into gate arrays implementing compounded logic operations.

In order to obtain all the desired input-output responses of the different gates, we need to satisfy the conditions enumerated in Table 4.1 simultaneously. So given a dynamics $f(x)$ corresponding to the physical device in actual implementation, one must find values of threshold and initial state satisfying the conditions derived from the truth tables to be implemented (see Table 4.2).

Table 4.1. *Truth table of the basic logic operations for a pair of inputs:* I_1, I_2 [1]*. The* 1*-input NOT gate is given by* NOT(0) *is* 1*;* NOT(1) *is* 0*.*

I_1	I_2	NAND	NOR	XOR	AND	OR	XNOR
0	0	1	1	0	0	0	1
0	1	1	0	1	0	1	0
1	0	1	0	1	0	1	0
1	1	0	0	0	1	1	1

Table 4.2. *Necessary and sufficient conditions, derived from the logic truth tables, to be satisfied simultaneously by the nonlinear dynamical element in order to have the capacity to implement the logical operations AND, OR, XOR, NAND, NOR, and NOT (cf. Table* 4.1*) with the same computing module* [5]*.*

Logic Operation	Input Set (I_1, I_2)	Output	Necessary and Sufficient Condition
AND	(0,0)	0	$f(x_0) \le x^*$
	(0,1)/(1,0)	0	$f(x_0 + I_{in}) \le x^*$
	(1,1)	1	$f(x_0 + 2I_{in}) - x^* = I_{in}$
OR	(0,0)	0	$f(x_0) \le x^*$
	(0,1)/(1,0)	1	$f(x_0 + I_{in}) - x^* = I_{in}$
	(1,1)	1	$f(x_0 + 2I_{in}) - x^* = I_{in}$
XOR	(0,0)	0	$f(x_0) \le x^*$
	(0,1)/(1,0)	1	$f(x_0 + I_{in}) - x^* = I_{in}$
	(1,1)	0	$f(x_0 + 2I_{in}) \le x^*$
NOR	(0,0)	1	$f(x_0) - x^* = I_{in}$
	(0,1)/(1,0)	0	$f(x_0 + I_{in}) \le x^*$
	(1,1)	0	$f(x_0 + 2I_{in}) \le x^*$
NAND	(0,0)	1	$f(x_0) - x^* = I_{in}$
	(0,1)/(1,0)	1	$f(x_0 + I_{in}) - x^* = I_{in}$
	(1,1)	0	$f(x_0 + 2I_{in}) \le x^*$
NOT	0	1	$f(x_0) - x^* = I_{in}$
	1	0	$f(x_0 + I_{in}) \le x^*$

A representative example is given in Table 4.3, which shows the exact solutions of the initial x_0 and threshold x^* which satisfy the conditions in Table 4.2 when the dynamical evolution is governed by the prototypical logistic equation

$$f(x) = 4x(1-x).$$

The constant $x^* = \frac{1}{4}$ is common to both input and output and to all logical gates.

Above we have explicitly shown how one can select out temporal responses corresponding to different logic gate patterns from a nonlinear system, and this ability allows us to construct flexible hardware. Contrast our use of nonlinear elements here with the possible use of linear systems on one hand and stochastic systems on the other. It is not possible to extract all the *different* logic responses from the *same* element in the case of

Table 4.3. *One specific set of solutions of the conditions in Table 4.2 which yield the logical operations AND, OR, XOR, NAND and NOT, with* $I_{in} = \frac{1}{4}$. *Note that these theoretical solutions have been fully verified in a discrete electrical circuit emulating a logistic map* [5].

Operation	AND	OR	XOR	NAND	NOT
x_0	0	1/8	1/4	3/8	1/2
x^*	3/4	11/16	3/4	11/16	3/4

linear components, as the temporal patterns are inherently very limited. So linear elements do not offer much flexibility or versatility. Stochastic elements, however, have many different temporal sequences. But they are *not deterministic*, so one cannot use them to *design* components. Only nonlinear dynamics enjoys both richness of temporal behavior as well as determinism.

Also note that while *nonlinearity* is absolutely necessary for implementing all the logic gates, chaos may not be always be necessary. In the representative example of the logistic map presented in Table 4.3, solutions for all the gates exist only at the fully chaotic limit of the logistic map. But the degree of nonlinearity necessary for obtaining all the desired logic responses will depend on the system at hand and on the specific scheme employed to obtain the input-output mapping. It may happen that certain nonlinear systems will allow a wide range of logic responses without actually being chaotic.

4.2.2 ▪ Logic from Continuous-Time Nonlinear Systems

We now present a method for obtaining logic responses from a continuous-time nonlinear system. Consider a continuous-time system described by the evolution equation $\dot{x} = F(x,t)$, where $\mathbf{x} = (x_1, x_2, \ldots x_N)$ are the state variables and $\mathbf{F}$ are nonlinear functions. In this system we choose a variable, say x_1, to be thresholded. Whenever the value of this variable exceeds a threshold x^*, it resets to x^*; i.e., when $x_1 > x^*$, then (and only then) $x_1 = x^*$.

Now the basic 2-input 1-output logic operation on a pair of inputs I_1, I_2 in this method simply involves the setting of an input-dependent threshold; namely the threshold is

$$x^* = I_C + I_1 + I_2,$$

where I_C is the dynamic control signal determining the functionality of the processor. By switching the value of I_C one can switch the logic operation being performed.

Again I_1 or I_2 has value 0 when logic input is 0 and has value I_{in} when logic input is 1. So the threshold x^* is equal to I_C when logic inputs are $(0,0)$, $I_C + I_{in}$ when logic inputs are $(0,1)$ or $(1,0)$, and $I_C + 2I_{in}$ when logic inputs are $(1,1)$.

The output is again interpreted as logic output 0 if $x_1 \leq x^*$, i.e., the excess above threshold is 0. The logic output is 1 if $x_1 > x^*$, and the excess above threshold $(x_1 - x^*) \sim I_{in}$.

Now for a NOR gate implementation ($I_C \equiv I_{NOR}$) the following must hold true (see truth table in Table 4.1):

(i) When input set is $(0,0)$, output is 1, which implies that for threshold $x^* = I_{NOR}$, we must have $(x_1 - x^*) \sim I_{in}$.

(ii) When input set is $(0,1)$ or $(1,0)$, output is 0, which implies that for threshold $x^* = I_{NOR} + I_{in}$, we must have $x_1 \leq x^*$.

(iii) When input set is $(1,1)$, output is 0, which again implies that for threshold $x^* = I_{NOR} + 2I_{in}$, we must have $x_1 \leq x^*$.

For a NAND gate ($I_C \equiv V_{NAND}$) the following must hold true (see truth table in Table 4.1):

(i) When input set is $(0,0)$, output is 1, which implies that for threshold $x^* = I_{NAND}$, we must have $(x_1 - x^*) \sim I_{in}$.

(ii) When input set is $(0,1)$ or $(1,0)$, output is 1, which implies that for threshold $x^* = I_{in} + I_{NAND}$, we must have $(x_1 - x^*) \sim I_{in}$.

(iii) When input set is $(1,1)$, output is 0, which implies that for threshold $x^* = I_{NAND} + 2I_{in}$, we must have $x_1 \leq x^*$.

One can find values of I_C that will satisfy all the above input-output associations in a robust and consistent manner, thereby yielding a gate that can morph between NOR and NAND characteristics [7].

4.2.3 ▪ Dynamic Logic from Nonlinear Evolution

Now we review a method for obtaining logic output from a nonlinear system using the time evolution of the state of the system. Namely, this concept uses the nonlinear characteristics of the time dependence of the state of the dynamical system to extract different responses from the system. The exciting advantage of this method is that a single nonlinear system can yield various logic operations, including combinational logic, at different time steps.

Again we have the inputs determining the initial state: $x_0 + I_1 + I_2$. Then the system evolves over n iterative time steps to updated state x_n. The evolved state is compared to a monitoring threshold x^*. If the state is greater than the threshold, one obtains logic output 1, and if the state is less than the threshold, one gets logic output 0. This process is repeated for subsequent iterations. (See Figure 4.1 for a representative example.)

Further, the ubiquitous combinational operation, bit-by-bit arithmetic addition (half-adder) involving two logic gate outputs, namely AND (to obtain carry) and XOR (to obtain first digit of sum), can be obtained directly in consecutive iterations, with a single one-dimensional chaotic element.

Table 4.4. *Necessary and sufficient conditions to be satisfied by a chaotic element in order to implement the logical operations NAND, AND, NOR, XOR, and OR during different iterations* [15].

LOGIC	NAND	AND	NOR	XOR	OR
Iteration n	1	2	3	4	5
Input (0,0)	$x_1 = f(x_0) > x^*$	$f(x_1) \leq x^*$	$f(x_2) > x^*$	$f(x_3) \leq x^*$	$f(x_4) \leq x^*$
Input (0,1)/(1,0)	$x_1 = f(x_0 + I_{in}) > x^*$	$f(x_1) \leq x^*$	$f(x_2) \leq x^*$	$f(x_3) > x^*$	$f(x_4) > x^*$
Input (1,1)	$x_1 = f(x_0 + 2I_{in}) \leq x^*$	$f(x_1) > x^*$	$f(x_2) \leq x^*$	$f(x_3) \leq x^*$	$f(x_4) > x^*$

Also, in conventional realization, the typical full-adder requires two half-adder circuits and an extra OR gate. So in total, the implementation of a full-adder requires 2 XOR gates, two AND gates and one OR gate. However, using the dynamical evolution of a

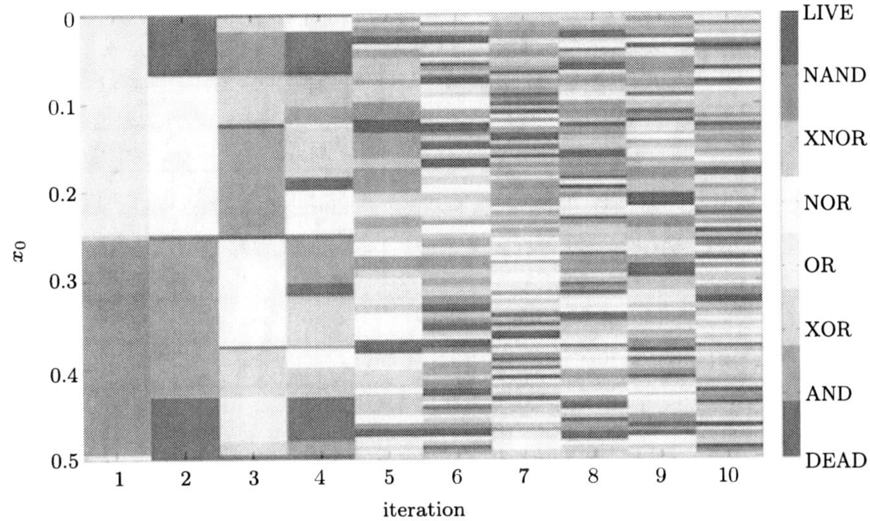

Figure 4.1. *Different logic patterns for varying* x_0 (0 *to* 0.5) *and iteration* n (0 *to* 10). *Here* $x^* = 0.75$ *for* $1 \leq n \leq 4$ *and* $x^* = 0.4$ *for* $n > 4$, *with* $I_{in} = 0.25$ [15].

single logistic map, we require only three iterations to implement the full-adder circuit directly, allowing combinational logic to be obtained very efficiently.

4.2.4 ▪ Using Synchronization to Obtain Dynamic Logic Gates

Now we review a scheme to obtain key logic gates using synchronization of nonlinear systems [12]. The idea will be explicitly demonstrated by numeric and experiments on nonlinear circuits. A significant feature of this scheme is that a single nonlinear drive-response circuit can be used to flexibly yield the different logic gates and switch logic behavior by small changes in the parameter of the response system. Namely, the response system can act as a "logic output controller."

Specifically, we outline the method for flexibly implementing the fundamental logic gate NOR (from which all gates may be constructed) and the logic gates AND and XOR (which yield the building block of arithmetic processing: the bit-by-bit addition operation). The logic element here is composed of one-way coupled nonlinear systems.

Inputs: The input set (I_1, I_2) determines the parameter of the drive system α_d: for example, input set $(0,0)$ can correspond to α_1, $(0,1)/(1,0)$ to α_2, and $(1,1)$ to α_3, where α_1, α_2, α_3 are three reasonably separated parameter settings of the drive system.

Output: The output is simply obtained from the synchronization error between the drive and response systems: large error is output 0, and output 1 corresponds to very small error. That is, *when the drive and response are quite synchronized, one obtains output* 1, *and* 0 *otherwise*.

Logic Control: The logic output is determined by the parameter setting of the response system; for example, if the parameter of the response system α_r is set close to α_1, one obtains NOR; $\alpha_r \sim \alpha_2$ gives XOR, and $\alpha_r \sim \alpha_3$ gives AND.

Further, other logic responses can also be obtained by using appropriate thresholds on the synchronization error to define output. The basic idea is to use *partial synchronization*

Table 4.5. *Updated state of chaotic element satisfying the conditions in Table 4.4 in order to implement the logical operations NAND, AND, NOR, XOR, and OR during different iterations with* $x_0 = 0.325$, $I_{in} = 1/4$, *and threshold* $x^* = 0.6$ [15].

Operation	NAND	AND	NOR	XOR	OR
Iteration n	1	2	3	4	5
State of the system (x_n)	x_1	x_2	x_3	x_4	x_5
Logic input(0,0) $x_0 = 0.325$	0.88	0.43	0.98	0.08	0.28
Logic input(0,1)/(1,0) $x_0 = 0.575$	0.9775	0.088	0.33	0.872	0.45
Logic input(1,1) $x_0 = 0.825$	0.58	0.98	0.1	0.34	0.9

Table 4.6. *Parameter of the drive system determined by the input set* I_1, I_2, *parameter of the response system giving logic function control; and synchronization error yielding the output of operation for the NOR, XOR, and AND logic gates* [12].

Logic Gate	Input Set (I_1, I_2)	Drive Parameter (Input Set)	Response Parameter (Logic Control)	Synchronization Error	Output
NOR	(0,0)	α_1	$\alpha_{NOR} \sim \alpha_1$	small	1
	(0,1)/(1,0)	α_2		large	0
	(1,1)	α_3		large	0
XOR	(0,0)	α_1	$\alpha_{XOR} \sim \alpha_2$	large	0
	(0,1)/(1,0)	α_2		small	1
	(1,1)	α_3		large	0
AND	(0,0)	α_1	$\alpha_{AND} \sim \alpha_3$	large	0
	(0,1)/(1,0)	α_2		large	0
	(1,1)	α_3		small	1

in order to generate a larger set of synchronization response patterns. For instance, to get NAND and OR logic one can use the logic control parameters in the response system α_{NAND} to lie between parameters α_1 and α_2 for the NAND gate characteristics, and a parameter setting α_{OR} lying in between α_2 and α_3 for OR gate characteristics.

Defining a suitable degree of synchronization, i.e., some small prescribed error, as output 1, and large synchronization error as output 0, we can get NAND and OR output responses as well. For instance, for NAND, if we design the parameters appropriately, partial synchronization will result in the first 2 input settings (as α_1 and α_2 are close to α_{NAND}) but not in inputs $I_1 = 1, I_2 = 1$ (whose drive parameter setting α_3 is far from α_{NAND}). Similarly for OR, only input set $I_1 = 0, I_2 = 0$ with parameter setting α_1 will give no synchronization (namely large error) and output 0, while the other two input sets (settings α_2 and α_3) will synchronize reasonably and give output 1 (as α_{OR} is in between, and close to, both α_2 and α_3).

Specifically, we have demonstrated this idea of obtaining key logic gates via synchronization in proof-of-principle experiments on nonlinear circuits (see Figures 4.3 and 4.4 for representative examples).

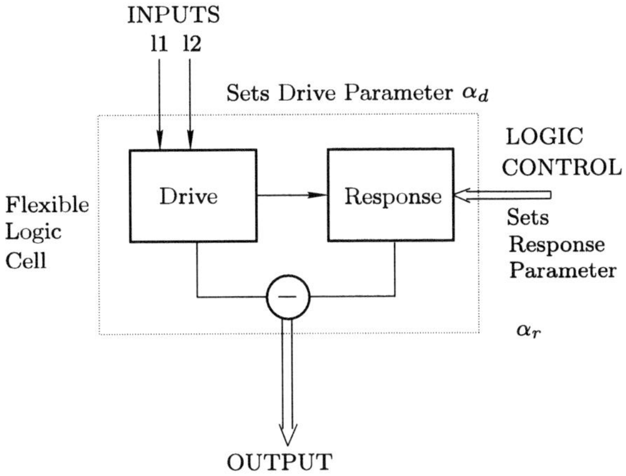

Figure 4.2. *Schematic diagram of the flexible dynamic logic cell comprised of one-way coupled nonlinear systems: the drive and the response, with the difference between the states of these determining the output* [12].

Note that it is also possible to *tune different parameters in parallel* in the drive for *multi-bit logic synthesis*. Such a realization will achieve parallel logic operations using the same logic cell.

4.3 ▪ Direct Implementation of SR Flip-Flop Using a Single Chaotic System

Now we describe the experimental implementation of the fundamental SR flip-flop (see Table 4.7) using a single chaotic Chua's circuit. Our approach directly obtains outputs from the system by simple threshold mechanisms, and does not necessitate the cross-coupling of NOR or NAND logic gates.

By utilizing a suitable threshold control mechanism, we show the direct implementation of SR flip-flop using a single chaotic circuit without requiring the cross-coupling of either NOR or NAND logic gates. Our logic processor is a continuous-time nonlinear system, described by the evolution equation $\dot{x} = F(x, t)$, where $x \equiv (x_1, x_2, x_3, \ldots, x_N)$ are the state variables and F are nonlinear functions. In this system, a variable x_T is chosen to be threshold controlled, as follows:

$$g(x_i) = x_i, \quad x_l^* < x_i < x_u^*, \tag{4.1}$$

$$g(x_i) = x_l^*, \quad x_i \leq x_l^*, \tag{4.2}$$

$$g(x_i) = x_u^*, \quad x_i \geq x_u^*, \tag{4.3}$$

with $g(x_T)$ being the output from the threshold controller. The threshold control will be triggered whenever the value of the monitored variable exceeds a critical upper threshold x_u^* or is less than a critical lower threshold x_l^*. The dynamics continues till the next occurrence of x_T exceeding (or dropping below) the threshold, when control resets its value to x_u^* (or x_l^*).

Now the basic logic operation on a pair of inputs I_1 and $-I_2$ in this method simply involves the setting of an input-dependent threshold; namely, the threshold levels x_u^* and x_l^* are determined by I_1 and $-I_2$.

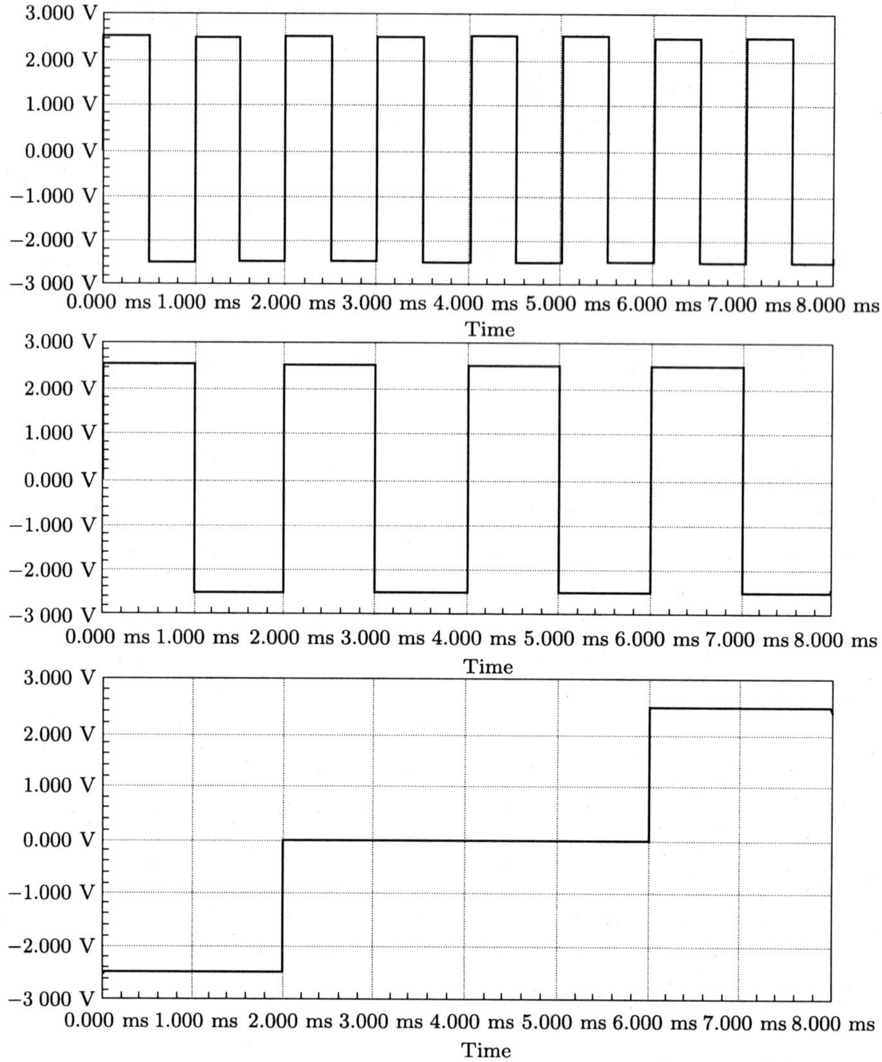

Figure 4.3. *Explicit circuit realization of the method with one-way coupled Chua's circuits* [12]*. The top two panels show a stream of input signals I_1 and I_2; the lowest panel shows the transmitted signal from drive to response* [12].

In our specific implementation, as the computing element, we consider a realization of the double scroll chaotic Chua's circuit represented by the following set of (rescaled) three coupled ODEs:

$$x_1 = \alpha(x_2 - x_1 - h(x_1)), \tag{4.4}$$

$$x_2 = x_1^* - x_2 + x_3, \tag{4.5}$$

$$x_3 = \beta x_2, \tag{4.6}$$

where $\alpha = 10$ and $\beta = 14.87$ and the piecewise linear function $h(x) = bx + 0.5(a-b)(|x+1| - |x-1|)$ with $a = -1.27$ and $b = -0.68$. Here $g(x)$ is the threshold controlled signal given by (4.3).

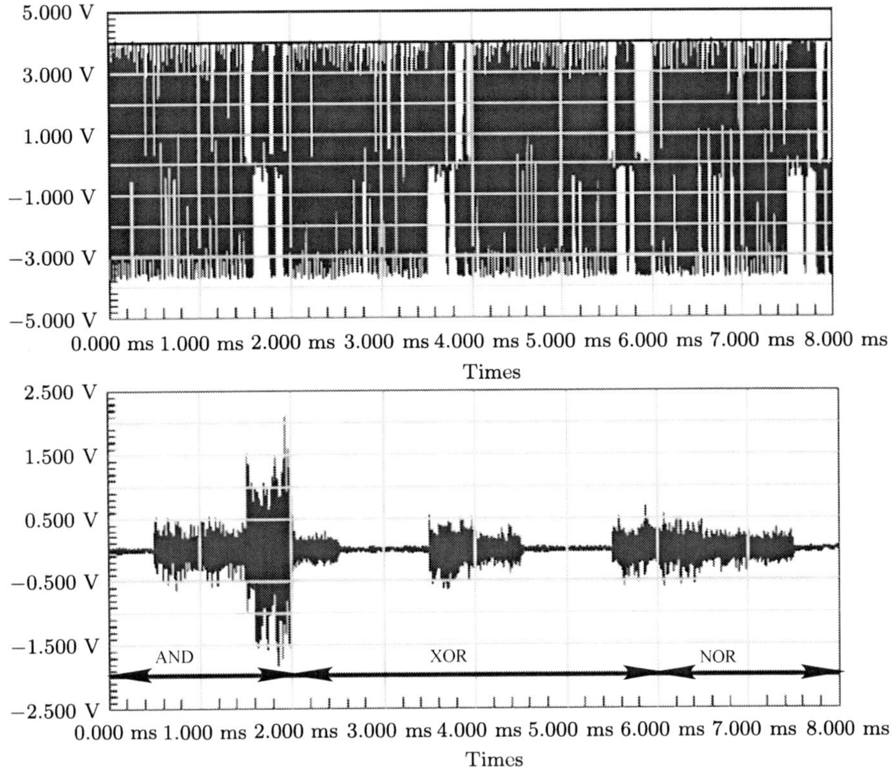

Figure 4.4. *Top panel shows the dynamic logic configuration signal; lower panel 2 shows the synchronization error signal giving a dynamic logic response:* 0–200 msec *is the AND response,* 200–600 msec *is the XOR response, and* 600–800 msec *is the NOR response. The signal transmitted from drive to response can be periodic or chaotic, depending upon the value of the tuning parameter in the drive system* [12].

Table 4.7. *Relationship between the two inputs and the output of SR latch.*

Set (I_1)	Reset (I_2)	Latch
0	0	No change (maintain the previous state)
0	1	0
1	0	1
1	1	Restricted Set

The schematic diagram for our method is depicted in Figure 4.5. Figure 4.6 shows the circuit diagram for implementing the controlling mechanism and for the realization of SR flip-flop. The circuit component values are fixed as L = 18 mH, R = 1710 Ohm, C1 = 10 nF, C2 = 100 nF, R1 = 220 Ohm, R2 = 220 Ohm, R3 = 2.2 kOhm, R4 = 22 kOhm, R5 = 22 Kohm, R6 = 3.3 kOhm, D = 1*N*4148, OA1-OA4 = μ A741 or AD712 opamps. Note that the circuit of Figure 4.5 is the ring structure configuration of the classic Chua's

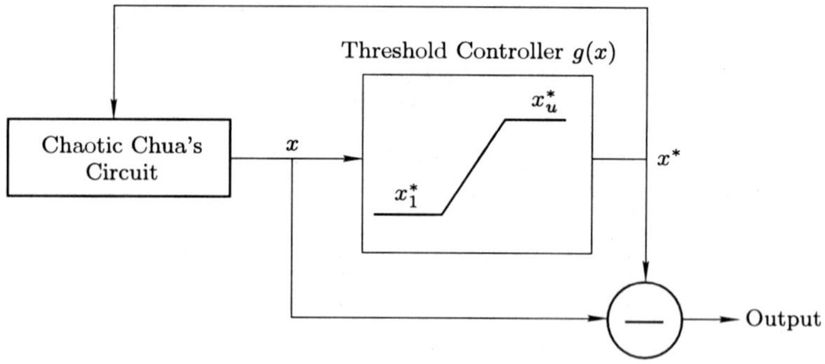

Figure 4.5. *Schematic diagram for implementing the SR flip-flop. The upper and lower threshold levels are controlled by the logic inputs SET and RESET, respectively.*

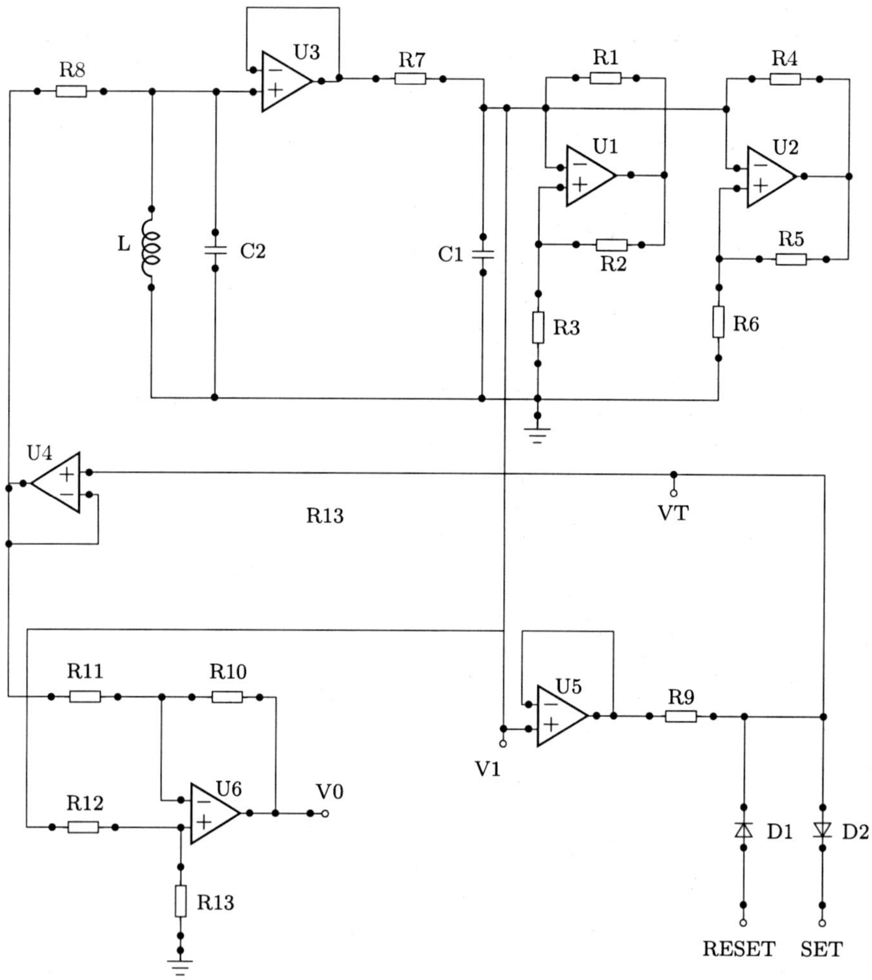

Figure 4.6. *Circuit module for implementing SR flip-flop. This circuit contains Chua's circuit with a threshold control circuit.*

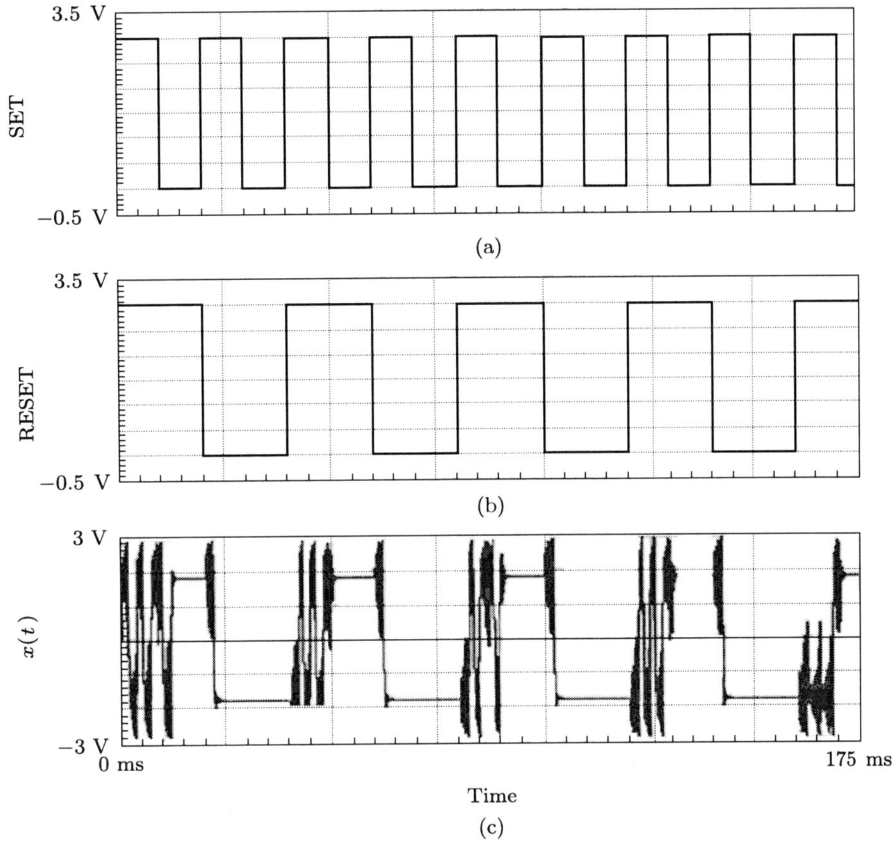

Figure 4.7. *SPICE simulations: Timing sequences from top to bottom:* (a) *input SET,* (b) *input RESET,* (c) *output* $x(t)$.

circuit by including the buffers U3 and U4. The buffer U5, resistor R9, and diodes D1 and D2 constitute the classic threshold controller. We use the x variable, corresponding to voltage V1, across capacitor C1 for thresholding. The input voltages to the diodes D1 and D2, which determine the upper and lower thresholds in (4.3), represent the SET and RESET logic inputs. A difference amplifier U6 gives the output V0, and VT is the signal from the threshold controller. In order to illustrate the dynamic operation of the circuit of Figure 4.6, consider the logic inputs SET and RESET waveforms as shown in Figure 4.7 applied to the diodes D2 and D1, respectively. The RESET logic input to diode D1 is applied via an inverting amplifier (not shown in Figure 4.6). Whenever SET voltage goes to 0V, the circuit output V1 (corresponding to state variable x_1 in (4.6)) goes to 1.84V (the positive fixed point state of (4.6)), whereas there is no control action if SET voltage goes to 3V. At the same time, if RESET voltage goes to 0V, then V1 goes to −1.84V (the negative fixed point state of (4.6)) whereas there is no control action if RESET voltage goes to −3V. Thus the dynamics of the circuit is controlled by the state of SET and RESET logic inputs, and by using suitable voltage comparators, the logic outputs Q and Qc satisfying the familiar SET-RESET flip-flop truth table can be obtained directly from the voltage waveform V0. The waveforms Q and Qc are shown in Figure 4.8.

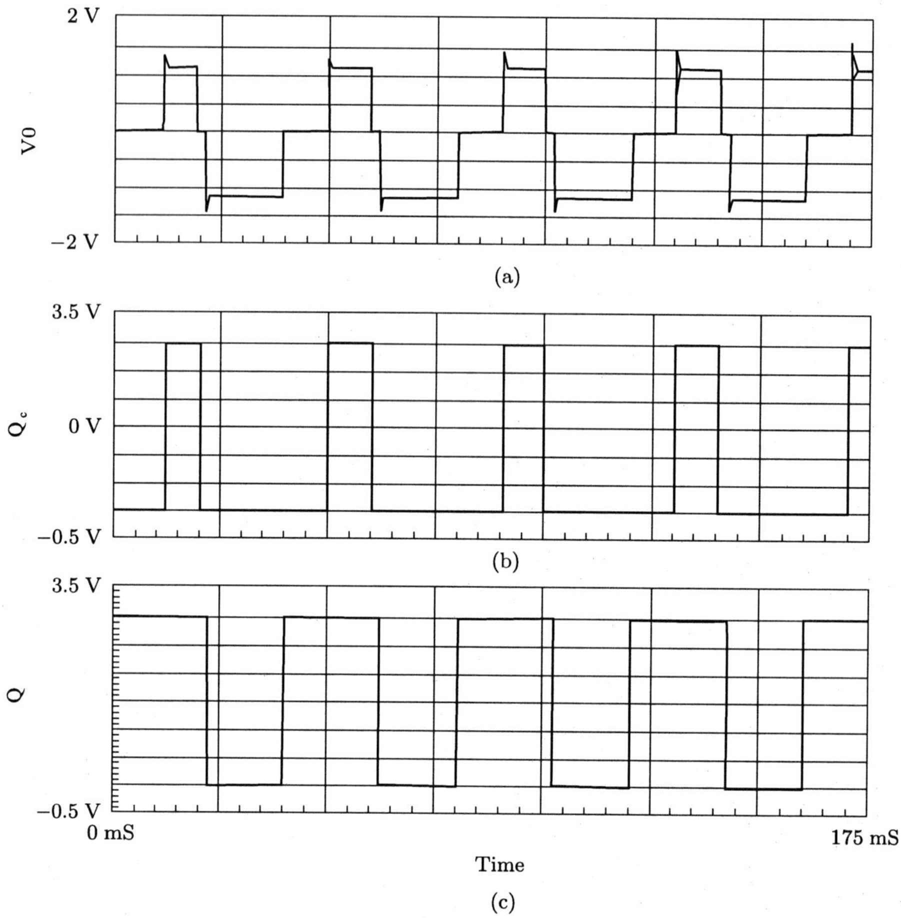

Figure 4.8. *SPICE simulations: Timing sequences:* (a) *Recovered output* $x(t) - x^*(t)$, (b) *output* Q, (c) *conjugate output* Qc.

Thus we have demonstrated how a simple threshold controlled chaotic Chua's circuit, subject to two logic inputs, can be used to implement the fundamental SR flip-flop directly, without requiring the cross coupling of either NOR or NAND logic gates.

4.4 ▪ Logical Cellular Automata

The basic cellular automaton (CA) [25, 26] is a D-dimensional lattice of identical cells, each of which is in one of a finite set of possible internal states (for instance, for a 2-state CA it is either 0 (OFF) or 1 (ON)). The entire lattice, which may be infinite, is updated synchronously in discrete time steps. On each update or iteration, each cell assumes a new state determined by its own present state, and the state of a certain set of adjacent or neighborhood cells set by a predefined rule. For example, in a one-dimensional CA, a cell is connected to r local cells on either side, where r is a parameter called the radius. Thus each cell has $2r + 1$ neighbors, including itself.

For a two-dimensional CAs, two types of cellular neighborhoods are usually considered: five cells, consisting of the cell along with its four immediate nondiagonal neighbors, and nine cells, consisting of the cell along with its eight surrounding neighbors. The update rule or transition function is identical for each cell, which specifies what internal state to assume for each possible arrangement of states in the adjacent cells; this is referred to as the "rule table" [25–28]. CAs have drawn wide attention and applicability in several research fields, including physics, biology, computer science, and so on. Specifically a growing interest in the utilization of CAs as computing devices has drawn considerable interest in areas including emergent computation [27], cellular structures for computing in the quantum regime [28], collision-based computing [29], and quantum CAs [30]. This is largely attributed to their massive parallelism, locality of cellular interactions, and simplicity of basic cells. Also CAs suggest a possible approach to formulate novel computational architectures at the nanometer scale [28] and cellular neural networks [10]. Also, in the majority of the schemes mentioned above, either genetic algorithms are used to evolve CAs to do particular computation, or array structure is used to do logic emulation [27]. In the present paper, we use CAs as the medium for emulating different logic gate operations in a very simple and straightforward way by interpreting the input and output in a particular manner. The presented scheme is generic in that all the different logic gate operations can be observed in a given CA with any specific local rule. There is a possibility of altering the local rule after certain updates or iterations of cells. Further, the applicability of this approach has been demonstrated by carrying-out certain simple bit-by-bit arithmetic operations. Now the following subsections are organized as follows: Subsection 4.4.1 describes the programming of the CAs; subsection 4.4.2 deals with the emulation of different logic operations with CAs. Then subsection 4.4.3 demonstrates the ability of to do a particular computing task, namely, bit-by-bit addition operation with CAs.

4.4.1 ▪ Programming the Cellular Automata

Consider CAs in a one-dimensional line of cells. Each cell C_i represents a dynamical system, discrete in time as well as space. If a grid is simply made up of discrete cells on a line, then this CA is 1D-CA. If we consider a plane containing $n \times n$ cells, then this CA is 2D-CA. At each site i of the lattice or grid there is a local variable u_i whose value is either 0 or 1. Each cell C_i at site i accepts three inputs, u_{i-1}, u_i, and u_{i+1}, where u_{i-1} denotes the input coming from the left neighboring cell C_{i-1}, u_i denotes the "self" input to cell C_i, and u_{i+1} denotes the input coming from the right neighboring cell C_{i+1}, as depicted in Figure 4.9. This type of dynamical interaction is a nearest neighbor interaction with radius of interaction $r = 1$. Thus, for $r = 1$, the value of the site i at the time $t+1$, is represented as

$$u_i(t+1) = f(u_{i-1}(t), u_i(t), u_{i+1}(t)), \tag{4.7}$$

where f is some Boolean function called a local rule or rule of interaction. The triplet $(u_{i-1}(t), u_i(t), u_{i+1}(t))$ takes its value from eight possible combinations as shown in Figure 4.9. The local rule f is uniquely defined when a value (0 or 1) is attributed to each triplet. Then a particular rule f encodes each triplet into a binary number. Thus each rule f yields a sequence of binary numbers which can be further identified by a decimal

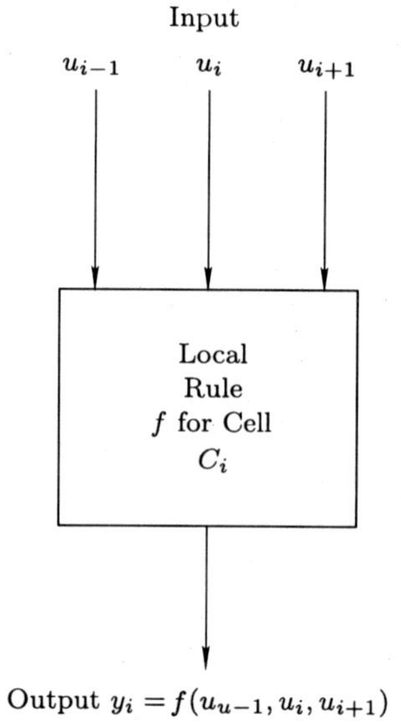

Figure 4.9. *Each cell C_i in the 1D cellular automata is described by a local rule N. The logical inputs are mapped as corresponding initial states u_{i-1}, u_i and u_{i+1}. Other cells are set at initial state 0.*

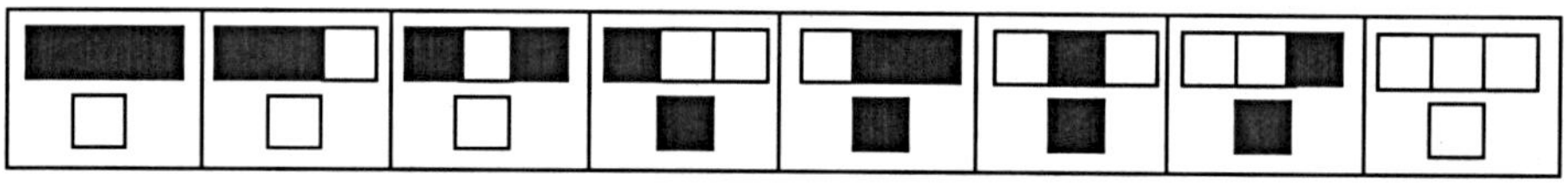

Figure 4.10. *A truth table representation of the CA "rule 30." The top row in each box gives one of the possible combinations of colors for a cell and its immediate neighbors. The connectivity radius is $r = 1$, meaning that each cell has two neighbors, one to its immediate left and one to its immediate right. Each cell can be in one of two states, denoted as 0 (open box) and 1 (solid box). The bottom row then specifies what color the center cell should be on the next iteration in each of these cases [9].*

number R [29]. One representative rule, "number 30" of 256 possible rules, is depicted in Figure 4.10 [26].

Thus, according to this rule, the one-dimensional CAs for given initial conditions at $t = 1$ evolve according to (4.7). After each time step (or update or iteration), only a single line of updated cells result. However, one can also consider a two-dimensional CA that involve a whole grid of cells, with the color of each cell being updated according to a local rule that depends on its neighbors in all four directions on the grid [25].

Now let us consider a 1D-CA with n number of cells. Now this CA receives an initial seed of logical 1 at site $i-1$, i, or $i+1$, which depends upon the state of logic input pair (I_1, I_2) for NOR, NAND, AND, OR, and XOR logic operations. In our scheme the action of the inputs is to fix the initial state ("excitation") of a particular cell to 1, keeping the rest of the cells at initial condition 0. With initial condition at a particular site i, determined by the input, the CA is iterated according to the local rule f. Upon single iteration or a set of consecutive iterations, the different cells exhibit different type of logic gate operations. The same scenario is applicable for a 2D-CA case also.

4.4.2 ▪ Emulation of Logic Gates

(a) Programming a 1D-CA. Here we will explicitly show, by using the 1D-CA, how to obtain clearly defined logic gate operations NOR, NAND, AND, OR, XOR (see Table 4.1 for the truth table). All the fundamental logic gate operations involve encoding the input by setting the initial value at a particular cell site (refer to Table 4.8).

(1) Inputs (for a 2 input operation): $C_i \to C_{prog}$, where $C_{prog} = C_i$ is the selected cell of the 1D-CA with a given grid size n. The initial value of this selected cell C_i and its neighbors is fixed as follows, depending upon the logic input pairs (I_1, I_2) condition:

Case 1. Both I_1, and I_2 are 0 (row 1 in Table 4.1), i.e., the initial state of the cell $C_{i-1} = 1$ and rest of the cells is set at initial condition $= 0$.

Case 2. $I_1 = 0$, $I_2 = 1$ (row 2 in Table 4.1), i.e., the initial state of the cell $C_i = 1$ and rest of the cells is set at initial condition $= 0$.

Case 3. $I_1 = 1, I_2 = 0$ (row 3 in Table 4.1), i.e., the initial state of the cell $C_i = 1$ and rest of the cells is set at initial condition $= 0$.

Case 4. Both $I_1 = I_2 = 1$ (row 4 in Table 4.1), i.e., initial state of the cell $C_{i+1} = 1$ and rest of the cells is set at initial condition $= 0$.

(2) Cell updates over some prescribed number of steps according to (4.7).

(3) The evolved cell state after certain iterations t yields the logic output as follows: Logic output $= 0$ if cell state is 0, Logic output $= 1$ if cell state is 1.

Table 4.8. *Symbolic truth table. Here u_i is input to the representative cell C_i, u_{i-1} is the input to the left neighbor cell C_{i-1}, and u_{i+1} is the input to the right neighbor cell C_{i+1} in the 1D-CA. Other cells have initial state at 0.*

Logic Input	u_{i-1}	u_i	u_{i+1}
(0,0)	1	0	0
(0,1)/(1,0)	0	1	0

In order to obtain all the desired input-output responses of the different gates, we need to satisfy the logic relationships enumerated in Table 4.1 for the state of the updated cells. As a representative result, the "rule 30" is adapted and the initial conditions are translated as in Table 4.8. Figures 4.11 and 4.12 depict the 1D-CA evolution for cell grid size $n = 20$ and iteration number $t = 10$. Figure 4.11 is generated for logic pair input condition $(I_1, I_2) = (0,0)$ with initial condition excited at cell number 9. Figure 4.12 is generated for logic pair input condition $(I_1, I_2) = (0,1)$ or $(1,0)$ with initial condition excited at cell number 10. Likewise, Figure 4.13 is generated for logic pair input condition $(I_1, I_2) = (1,1)$, for which the initial condition is excited at cell number 1_1.

It can be noticed that the local "rule 30" is uniformly used for all these cases. Then by interpreting the results for logic gate outputs, Figure 4.14 is generated by identifying the logic states of the corresponding cells and is appropriately color coded. Specifically, for instance, to determine the logic of cell $i = 9$ in Figure 4.14, we have to inspect Figures 4.11–4.13. Figure 4.11 shows that cell $i = 9$ is black (output 1) after 1 iteration when the cell $i = 9$ is "excited" to 1; namely input is $(0,0)$. Figure 4.12 shows, after 1 iteration, cell $i = 9$ is black (output 1) when cell $i = 10$ is "excited" to 1 namely for input sets $(0,1)$ or $(1,0)$.

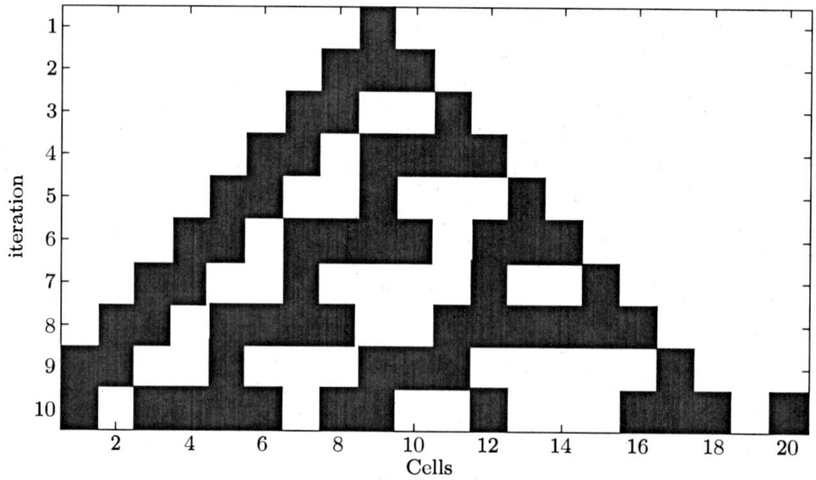

Figure 4.11. *A 1D-CA with "rule 30" corresponding to truth table of Figure 4.10. The CA is started from a single black cell or seed. The cell grid size is* $n = 20$*, and the first 10 iterations or updates are shown. The initial seed at cell 9 is excited to logic 1 (solid box) for the logic input pair* $(I_1, I_2) = (0,0)$*. The initial condition for rest of the cells is set at logic 0 (open box).*

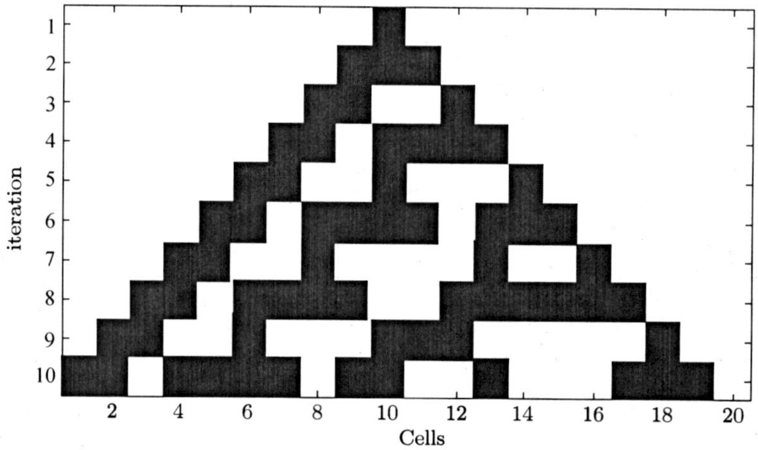

Figure 4.12. *A 1D-CA with "rule 30." The initial seed at cell 10 is excited to logic 1 (solid box) for the logic input pair* $(I_1, I_2) = (0,1)$ *or* $(1,0)$*. The initial condition for rest of the cells is set at logic 0 (open box).*

Last, cell $i = 9$ is white (output 0) for cell $i = 1_1$ being 1, namely input set $(1,1)$. So this cell after 1 iteration (i.e., $t = 1$, namely the first row in Figure 4.14) corresponds to logic type NOR (see truth table for input-output association). Similarly (by inspecting Figures 4.11–4.13) one can straightforwardly find the logic operation emulated by the other cells after a particular iteration. For example, one can identify the logic gate output for cells 3, 8, and 20 as logic XNOR operation for iteration or update value $t = 10$.

Likewise, different cells at different iterations or updates emulate logic operations as color coded in Figure 4.14. Thus, the spatial position of the excitation (i.e., initial state 1 in a grid of cells at 0) is determined by the inputs. The state of the cell and its neighbors

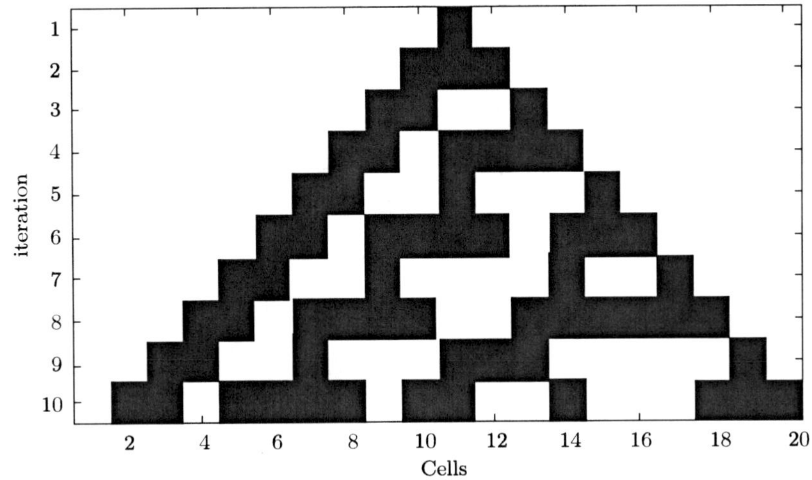

Figure 4.13. *A 1D-CA with "rule 30." The initial seed at cell 11 is excited to logic 1 (solid box) for the logic input pair* $(I_1, I_2) = (1,1)$. *The initial condition for rest of the cells is set at logic 0 (open box).*

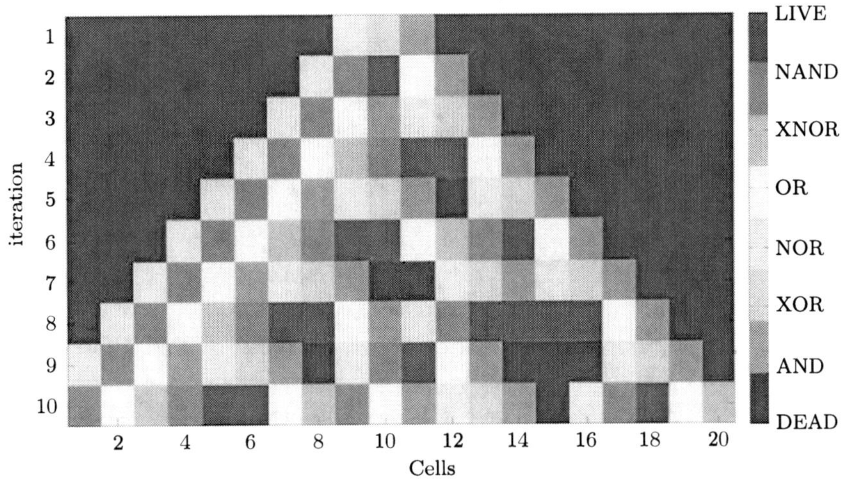

Figure 4.14. *Template showing different logic gate operations exhibited by the cells for different iterations or updates. This figure is generated by interpreting the cell states of Figures 4.11–4.13. For example, cell 3, cell 8, and cell 20 depict logic gate XNOR operation in 10 iterations.*

after a specified number of iterations yields the output. Therefore, different kinds of logic responses "propagate" around the reference input cell. This interpretation of CAs in terms of logic type is suggested for the first time here.

(b) Programming 2D-CA $C_{i,j} \to C_{prog}$, where $C_{prog} = C_{i,j}$ is the selected cell of the 2D-CA with a given grid size (n,n). The initial value of this selected cell $C_{i,j}$ and its neighbors are fixed as follows, depending upon the logic input pairs (I_1, I_2) condition:

Case 1. Both I_1, and I_2 are 0 (row 1 in Table 4.1), i.e., the initial state of the cell $C_{i,j} = 1$ and rest of the cells is set at initial condition = 0.

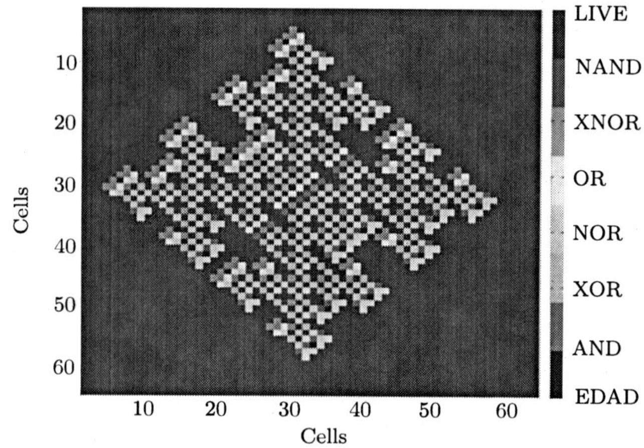

Figure 4.15. *Template showing different logic gate operations exhibited by the cells for different iterations or updates with 2D-CA. The CA rule code is "942." A two input logic* (I_1, I_2) *configuration is considered. The cell grid size is* $(64, 64)$. *Here cell* $(32, 32)$ *is excited to logic* 1 *if logic inputs* $(I_1, I_2) = (0, 0)$. *Cell* $(31, 31)$ *is excited to logic* 1 *if logic inputs* $(I_1, I_2) = (0, 1)$ *or* $(1, 0)$. *Cell* $(30, 30)$ *is excited to logic* 1 *if logic inputs* $(I_1, I_2) = (1, 1)$. *For example, cell* $(28, 53)$, *cell* $(28, 54)$, *cell* $(31, 57)$, *cell* $(32, 58)$, *and cell* $(31, 56)$ *exhibit AND* (I_1, I_2), *OR* (I_1, I_2), *NAND* (I_1, I_2), *NOR* (I_1, I_2), *and XOR* (I_1, I_2), *respectively, for iteration* = 15.

Case 2. $I_1 = 0, I_2 = 1$ (row 2 in Table 4.1), i.e., the initial state of the cell $C_{i-1,j-1} = 1$ and rest of the cells is set at initial condition = 0.

Case 3. $I_1 = 1, I_2 = 0$ (row 3 in Table 4.1), i.e., the initial state of the cell $C_{i-1,,j-1} = 1$ and rest of the cells is set at initial condition = 0.

Case 4. Both I_1 and I_2 are 1 (row 4 in Table 4.1), i.e., initial state of the cell $C_{i-2,j-2} = 1$ and rest of the cells is set at initial condition = 0.

(4) Cell updates over some prescribed number of steps according to a given local rule.

(5) The evolved cell state after certain iterations t yields the logic output as follows: logic output = 0 if cell state is 0, logic output = 1 if cell state is 1. Figure 4.15 exhibits the color coded diagram for different logic operations being revealed by different cells after iteration time $t = 15$ and $C_{i,j} = (30, 30)$.

4.4.3 ▪ Implementation of Bit-by-Bit Arithmetic Functions

We now demonstrate how one can obtain the ubiquitous bit-by-bit arithmetic addition using 1D-CA with $n = 20$ cells and local rule 30. A simple 1-bit binary arithmetic addition requires a full adder logic, which adds three individual bits together (two bits being the digit inputs and the third bit assumed to be carry from the addition of the next least-significant bit addition operation, known as C_{in}).

A typical full-adder requires two half-adder circuits and an extra XOR gate. In total, the implementation of a full-adder requires five different gates (three XOR gates and two AND gates). However, in the present implementation of utilizing the dynamical evolution of a 1D-CA we need only consecutive iterations or updates for a selected cell to emulate the full-adder logic. Now we will employ three steps to implement our full-adder logical operations:

(1) Inputs (for a 3-input operation) $C_i \rightarrow C_{prog}$, where $C_{prog} = C_i$ is the selected cell of the 1D-CA with a given grid size n. The initial value of this selected cell C_i and its neighbors is fixed as follows, depending upon the logic inputs (I_1, I_2, I_3) condition:

Case 1. All inputs are 0 (row 1 in Table 4.9), i.e., the initial state of the cell $C_{i-1} = 1$ and rest of the cells are set at initial condition $= 0$.

Case 2. One of the input $= 1$ (rows 2, 3, and 5 in Table 4.9), i.e., the initial state of the cell $C_i = 1$ and rest of the cells is set at initial condition $= 0$.

Case 3. Any two inputs $= 1$ (rows 4, 6, and 7 in Table 4.9), i.e., the initial state of the cell $C_{i+1} = 1$ and rest of the cells is set at initial condition $= 0$.

Case 4. All inputs $= 1$ (row 8 in Table 4.9), i.e., the initial state of the cell $C_{i+2} = 1$ and rest of the cells is set at initial condition $= 0$.

Table 4.9. *The truth table of full-adder.*

Input Bit for No. A	Input Bit for No. B	Carry Bit Input C_{in}	Carry Bit Output C_{out}	Sum Bit Output S
0	0	0	0	0
0	0	1	0	1
0	1	0	0	1
0	1	1	1	0
1	0	0	0	1
1	0	1	1	0
1	1	0	1	0
1	1	1	1	1

(2) Cell updates over some prescribed number of steps according to (4.7).

(3) The evolved cell state after certain iterations t yields the logic output as follows: logic output $= 0$ if cell state is 0, logic output $= 1$ if cell state is 1.

A representative plot of 1D-CA for full-adder logic operations is shown in Figure 4.16 for $n = 26$ cells, $t = 13$, and $i = 12$. The cell $i = 14$ at iteration 10 exhibits C_{out} logic states, and cell $i = 14$ at iteration 12 exhibits the sum bit output S.

Interestingly, one can observe from Figure 4.16 that different cells reveal other combinatorial logic operations for the 3-input case. For example, cell $i = 24$ depicts the logic NOR (I_1, I_2, I_3) operation for iteration (or update) value is $t = 13$. Cell 23 exhibits the logic AND (I_1, I_2, I_3) operation for iteration (or update) value $t = 13$. Cell 21 exhibits the logic XOR (I_1, I_2, I_3) operation for iteration 13, cell 20 exhibits the logic XNOR (I_1, I_2, I_3) operation for iteration 1_2, and cell 22 exhibits the logic NAND (I_1, I_2, I_3) operation for iteration 1_2. It is important to note that the actual computational overhead in this approach is minimal, as we are taking advantage of the natural evolution and behavioral patterns of CAs. This approach is not restricted to a particular local rule for CA evolution of identification of logic operations. However, in some cases for certain local rules, we cannot extract the desired or all basic logic operations from CA evolution. In these cases, some local rules result in either no evolution or some simple CA behaviors.

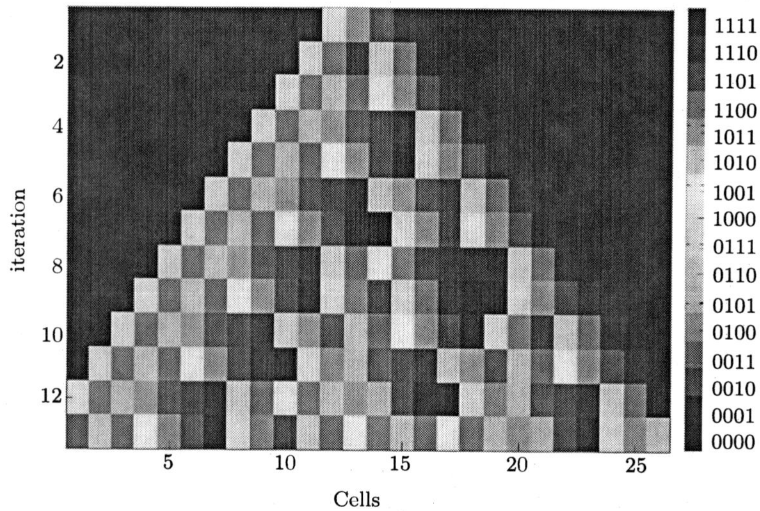

Figure 4.16. *Template showing different logic gate operations exhibited by the cells for different iterations or updates with 2D-CA. The CA rule code is "942." A two input logic* (I_1, I_2) *configuration is considered. The cell grid size is* $(64, 64)$. *Here cell* $(32, 32)$ *is excited to logic* 1 *if logic inputs* $(I_1, I_2) = (0, 0)$. *Cell* $(31, 31)$ *is excited to logic* 1 *if logic inputs* $(I_1, I_2) = (0, 1)$ *or* $(1, 0)$. *Cell* $(30, 30)$ *is excited to logic* 1 *if logic inputs* $(I_1, I_2) = (1, 1)$. *For example, cell* $(28, 53)$, *cell* $(28, 54)$, *cell* $(31, 57)$, *cell* $(32, 58)$, *and cell* $(31, 56)$ *exhibit AND* (I_1, I_2), *OR* (I_1, I_2), *NAND* (I_1, I_2), *NOR* (I_1, I_2), *and XOR* (I_1, I_2), *respectively, for iteration* $= 15$.

In summary, we have demonstrated the basic principles for the direct and flexible implementation of basic logic operations utilizing one-dimensional and two-dimensional CAs. By exciting the initial condition of a selected cell that depends upon the logic input state, subsequent iterations of CAs reveal different types of fundamental logical operations. The richness of the dynamics of CAs allows us to observe different logic gate responses from a selected cell by simply setting a suitable local rule and iteration. So different kinds of logic responses "propagate" around the reference input cell. This interpretation of CAs is suggested for the first time here. Further, we have shown how the dynamics of cellular automata can be utilized to emulate bit-by-bit arithmetic operations and combinatorial logic responses for 3-input logic gates. Interestingly, in order to implement logic involving n distinct input conditions we have to use n cells. The output can be obtained in a few (often one) iterations. Thus, multiple input combinational logic involves a spatial cost but no time overheads.

4.5 ▪ Summary

In conclusion, we have demonstrated the direct and flexible implementation of all the basic logic gates utilizing nonlinear dynamics. The richness of the dynamics allows us to select all the different gate responses from the same processor by simply setting suitable threshold levels. These threshold levels are known exactly from theory and thus available as a lookup table. Arrays of such logic gates can conceivably be programmed on the run (for instance, with a stream of threshold values being sent in by an external program) to be optimized for the task at hand. For instance such a morphing device may serve flexibly as an arithmetic processing unit or an unit of memory and can be swapped, as the need demands, to be one or the other. Thus architectures based on such logic implementations may serve as ingredients of a general-purpose reconfigurable computing device more powerful and fault tolerant than statically wired hardware.

References

[1] M. M. Mano, *Computer System Architecture*, 3rd ed., Prentice Hall, Englewood Cliffs, 1993; T. C. Bartee, *Computer Architecture and Logic Design*, McGraw-Hill, New York, 1991.

[2] S. Sinha and W. L. Ditto, "Dynamics based computation," *Phys. Rev. Lett.*, 81, pp. 2156–2159, 1998.

[3] S. Sinha, T. Munakata, and W. L. Ditto, "Parallel computing with extended dynamical systems," *Phys. Rev. E*, 65, 036214, 2002; T. Munakata, S. Sinha, and W. L. Ditto, "Chaos computing: Implementaion of fundamental logical gates by chaotic elements," *IEEE Trans. Circuits Syst. I*, 49, pp. 1629–1633, 2002; T. Munakata and S. Sinha, "Implementation of fundamental logical gates by 1-D chaotic elements," *Proc. COOL Chips VI*, Yokohama, p. 73, 2003.

[4] S. Sinha and W. L. Ditto, "Computing with distributed chaos," *Phys. Rev. E*, 60, pp. 363–377, 1999; S. Sinha, T. Munakata, and W. L. Ditto, "Flexible parallel implementation of logic gates using chaotic elements," *Phys. Rev. E*, 65, 036216, 2002; W. L. Ditto, K. Murali, and S. Sinha, "Exploiting chaos for computation," *Proceedings of IEEE Asia-Pacific Conference on Circuits and Systems (APCCAS06)*, Singapore, pp. 1835–1838, 2006.

[5] K. Murali, S. Sinha, and W. L. Ditto, "Construction of a reconfigurable dynamic logic cell," *Pramana*, 64, pp. 433–441, 2005.

[6] K. Murali, S. Sinha, and W. L. Ditto, "Implementation of NOR gate by a chaotic Chua's circuit," *Internat. J. Bifur. Chaos Appl. Sci. Engrg.*, 13, pp. 2669–2672, 2003; K. Murali, S. Sinha, and I. R. Raja Mohamed, "Chaos computing: Experimental realization of NOR gate using a simple chaotic circuit," *Phys. Lett. A*, 339, pp. 39–44, 2005.

[7] K. Murali, S. Sinha, and W. L. Ditto, "Chaos computing: ideas and implementation," *Philos. Trans. R. Soc. Lond. Ser. A*, 366, 2007; K. Murali, S. Sinha, and W. L. Ditto, "Reconfigurable logic element using a chaotic circuit," *Proceedings of IEEE Asia-Pacific Conference on Circuits and Systems (APCCAS06)*, Singapore, pp. 1839–1842, 2006.

[8] B. S. Prusha and J. F. Lindner, "Nonlinearity and computation: Implementing logic as a nonlinear dynamical system," *Phys. Lett. A*, 263, pp. 105–111, 1999.

[9] D. Cafagna and G. Grassi, "Chaos-based computation via Chua's circuit: Parallel computin with application to the SR flip-flop," *Internat. Symposium on Signals, Circuits and Systems*, 2, pp. 749–752, 2005.

[10] K. E. Chlouverakis and M. J. Adams, "Optoelectronic realization of NOR logic gate using chaotic two-section lasers," *Electronics Letters*, 41, pp. 359–360, 2005.

[11] M. R. Jahed-Motlagh, B. Kia, W. L. Ditto, and S. Sinha, "Fault tolerance and detection in chaotic computers," *Internat. J. Bifur. Chaos Appl. Sci. Engrg.*, 17, p. 1955, 2007.

[12] K. Murali and S. Sinha, "Using synchronization to obtain dynamic logic gates," *Phys. Rev. E*, 025201, 2007.

[13] A. Miliotis, K. Murali, S. Sinha, W. L. Ditto, and M. L. Spano, "A simple nonlinear dynamical computing device," *Chaos Solitons Fractals*, 42, pp. 809–819, 2009.

[14] A. Miliotis, S. Sinha, and W. L. Ditto, "Exploiting nonlinear dynamics to store and process information," *Internat. J. Bifur. Chaos Appl. Sci. Engrg.*, 18, pp. 1551–1559, 2008; A. Miliotis, S. Sinha, and W. L. Ditto, "Exploiting nonlinear dynamics to search for the existence of matches in a database," *Proceedings of IEEE Asia-Pacific Conference on Circuits and Systems (APCCAS06)*, Singapore, pp. 1843–1846, 2006.

[15] K. Murali, A. Miliotis, W. L. Ditto, and S. Sinha, "Logic from nonlinear dynamical evolution," *Phys. Lett. A*, pp. 1346–1351, 2009.

[16] K. Murali, S. Sinha, W. L. Ditto, and A. R. Bulsara, "Reliable logic circuit elements that exploit nonlinearity in the presence of a noise floor," *Phys. Rev. Lett.*, 102, 104101, 2009.

[17] W. L. Ditto, S. Sinha, and K. Murali, *Method and apparatus for a chaotic computing module using threshold reference signal implementation*, US patent 07096347, August 22, 2006.

[18] W. L. Ditto, A. Miliotis, K. Murali, S. Sinha, and M. L. Spano, "Chaogates: Morphing logic gates that exploit dynamical patterns," *Chaos*, 20, 037107, 2010.

[19] G. Taubes, "Computer design meets Darwin," *Science*, 277, pp. 1931–1932, 1997.

[20] S. Sinha and D. Biswas, "Adaptive dynamics on a chaotic lattice," *Phys. Rev. Lett.*, 71, pp. 2010–2013, 1993; L. Glass and W. Zheng, "Bifurcations in flat-topped maps and the control of cardiac chaos," *Internat. J. Bifur. Chaos Appl. Sci. Engrg.*, 4, p. 1061, 1994; S. Sinha, "Unidirectional adaptive dynamics," *Phys. Rev. E*, 49, pp. 4832–4842, 1994; S. Sinha, "Adaptive dynamics on circle maps," *Phys. Lett. A*, 199, pp. 365–374, 1995; S. Sinha and W. L. Ditto, "Controlling neuronal spikes," *Phys. Rev. E*, 63, 056209, 2001; S. Sinha, "Using thresholding at varying intervals to obtain different temporal patterns," *Phys. Rev. E*, 63, 036212, 2001; S. Sinha, "Controlling chaos with threshold mechanisms," *Nonlinear Systems*, R. Sahadevan and M. L. Lakshmanan (eds.), Narosa, pp. 309–328, 2002; W. L. Ditto and S. Sinha, "Exploiting the controlled responses of chaotic elements to design configurable hardward," *Phil. Trans. R. Soc. Lond. Ser. A*, pp. 2483–2494, 2006.

[21] K. Murali and S. Sinha, "Experimental realization of chaos control by thresholding," *Phys. Rev. E*, 68, 016210, 2003.

[22] R. J. Maddock and D. M. Calcutt, *Electronics: A Course for Engineers*, Addison Wesley Longman Ltd., London, 1997.

[23] M. Lakshmanan and K. Murali, *Chaos in Nonlinear Oscillators: Controlling and Synchronization*, World Scientific, Singapore, 1996.

[24] G. Q. Zhong, R. Bargar, and K. S. Halle, "Circuits for voltage tuning the parameters of Chua's circuit: Experimental application for musical signal generation," *J. Franklin Inst.*, 333, pp. 743–784, 1994.

[25] S. Wolfram, *A New Kind of Science*, Wolfram Media, 2002.

[26] L. O. Chua, *A Nonlinear Dynamics Perspective of Wolfram's New Kind of Science*, World Scientific, Singapore, 2007.

[27] M. Sipper, "Computing with cellular automata: Three cases for nonuniformity," *Phys. Rev. E*, 57, pp. 3589–3592, 1998; M. Sipper and M. Tamassini, "Computation in artificially evolved, non-uniform cellular automata," *Theoret. Comput. Sci.*, 217, pp. 81–98, 1999.

[28] S. C. Benjamin and N. F. Johnson, "A possible nanometer-scale computing device based on an adding cellular automaton," *Appl. Phys. Lett.*, 70, pp. 2321–2323, 1997; S. C. Benjamin and N. F. Johnson, "Cellular structures for computation in the quantum regime," *Phys. Rev. A*, 60, pp. 4334–4337, 1999.

[29] A. Adamatzky, *Computing in Nonlinear Media and Automata Collectives*, IOP, Bristol, 2001.

[30] I. Amlani, A. O. Orlov, G. Toth, G. H. Bernstein, C. S. Lent, and G. L. Snider, "Digital logic gate using quantum-dot cellular automata," *Science*, pp. 289–291, 1999.

Chapter 5

System Identification Using Chaos

Henry Leung and Ajeesh Kurian

Abstract: In this chapter, system identification using chaotic sequences is considered. Chaotic sequences are well suited for system identification since they possess wideband spectra and satisfy the persistent excitation condition. In addition, they are deterministic in nature and provide extra information for identification. We consider here identifying the linear moving average system and utilize the expectation maximization algorithm to perform parameter estimation. The sufficient statistics in the expectation step (E-step) is obtained with an unscented Kalman smoother (UKS). In order to assess the performance of the estimator, we derive the Cramér–Rao lower bound (CRLB) of the linear moving average parameter estimation problem when the driving sequences are chaotic. We show that blind identification of system driven by chaotic sequences has performance close to nonblind identification of the same using random sequences. In addition, the asymptotic error of the proposed method approaches the CRLB. The estimation performance of the proposed algorithm is then evaluated using computer simulations and compared against statistical system identification schemes. We further extend this to system identification using chaotic symbolic sequence. The improved identification based on chaotic symbolic sequence is also demonstrated. In this chapter, we use two-way ranging as an application of the proposed chaotic identification technique.

5.1 ▪ Introduction

In many practical engineering applications, there are instances where parameters of finite impulse response (FIR) systems have to be identified, such as channel estimation in communication systems [1]. In most of these situations, identification of such a system is an intermediate step, and the overall system performance depends on the accuracy of the estimated parameters. For example, in digital communication systems, channel estimation may be introduced as an intermediate step prior to channel equalization and symbol detection [2]. Training sequence–based nonblind identification is one of the most popular methods used in system identification. It requires precise knowledge of the training sequences and a perfect synchronization of the system input and output [3]. It is also less efficient since a part of the bandwidth is allocated to the training sequence which has to be transmitted periodically to track the changing channel conditions. Blind system identification has attracted significant interests since it does not require much prior knowledge about the input [2]. It is based on tactical properties of the input to the system. Semiblind identification schemes fall between the nonblind and blind identification, where some prior knowledge of the signal such as the shape of the pulse shaping waveform or binary pattern are available for system identification [1].

Chaotic sequences have found many applications in control, cryptography, and communications because of its noiselike appearance and wideband spectrum characteristics. Chaos has been found to be an effective way to drive linear systems for identification [3–5]. However, the potential of chaotic sequences for system identification still remains unexplored to a large extent. It is probably due to the sensitive dependence of chaotic systems on their initial conditions and the difficulty in synchronizing noisy chaotic signals [6]. In this chapter, we evaluate the application of chaotic signals as driving sequences for blind or semiblind system identification. We also investigate the use of chaotic numeric sequence and symbolic sequence [7] in system identification.

We formulate the system identification problem using maximum likelihood estimation (MLE), and the estimation is carried out by the expectation maximization (EM) algorithm [8]. Since the input sequence dynamics is essentially nonlinear, in the expectation step, we use an unscented Kalman smoother (UKS) to obtain the sufficient statistics. Although the proposed estimator is not restricted to any particular class of systems, we consider a generic FIR system in this chapter. We compare the performance of this chaos identification scheme with blind identifications methods based on higher-order statistics [1] and schemes where the knowledge about the input sequence is available at the receiver side (nonblind estimators).

Ranging is an important operation in military as well in civilian applications. Among many techniques used, pseudorandom binary sequence–based ranging has attracted considerable research attention [9, 10]. One of the main complications in such systems is the difficulty in synchronizing the incoming sequence when there are multiple copies of the transmitted signal due to multipath fading. A noncoherent estimation and detection is therefore more suitable in this case. We apply the proposed chaos identification approach to ranging by building a noncoherent range estimator so that the channel and transmitted signal can be estimated simultaneously. It is shown here that the proposed scheme is quite robust to the multipath fading effects.

The rest of this chapter is organized as follows. The maximum likelihood formulation of the chaos-based system identification problem is presented in section 5.2. Some popular blind identification methods in the literature are also introduced. In section 5.3, we evaluate the identification performance using computer simulations and derive the Cramér–Rao lower bound (CRLB) for theoretical assessment. As an application of the chaotic identification method, a noncoherent ranging scheme is developed here, and the results are discussed in section 5.4. Concluding remarks are given in section 5.5.

5.2 ▪ Problem Formulation

Perturbation signal design is one of the key tasks in system identification [11]. It is desirable to have an impulselike autocorrelation, which has a flat spectrum in the frequency domain. Hence, the system under study is persistently excited when it is fed with such an input signal. One of the popular choices that satisfy this condition is filtered Gaussian noise [10]. Another important characteristic of an input signal is its crest factor. A low crest factor is desired since it will result in maximum total energy to the system. Random binary sequences, especially pseudorandom binary sequence are widely used as input signals for system identification. Recently, signals generated from chaotic dynamics have been investigated for system identification. Chaotic signals are characterized by their noiselike appearance and wide spectra. Another advantage over white Gaussian noise is that chaotic signals can be generated by simple dynamic systems that are relatively easy to implement. We consider here a FIR system excited by a chaotic signal z_n. The output of the system which is corrupted by noise can be expressed by

$$y_n = \mathbf{h}^T \mathbf{z}_n + w_n, \tag{5.1}$$

where $\mathbf{z}_n = [z_n, \ldots, z_{n-p+1}]$ is the input to the system with coefficients $\mathbf{h} = [h_1, \ldots, h_p]$ and w_n is zero mean additive white Gaussian noise with variance σ_w^2. A chaotic signal $c_n, n = 0, \ldots, N-1$, where N is the total of number of points, can be generated by a one-dimensional nonlinear map,

$$c_{n+1} = f(c_n, \eta), \tag{5.2}$$

where η is the bifurcating parameter of the chaotic map. Using c_n as the driving signal to identify the system (5.1), we simply set $z_n = c_n$. If a binary symbolic sequence is to be used to drive the system in (5.1), we can obtain the symbolic chaotic signal s_n from

$$s_n = g(c_n) = \begin{cases} +1 & \text{if } c_n \geq \mu, \\ -1 & \text{otherwise.} \end{cases} \tag{5.3}$$

Blind identification is used to estimate $\mathbf{h}$ from the observations y_n without using the input sequence. The estimation problem becomes estimating $\mathbf{h}$ and σ_{w}^2 in addition to the initial condition c_0. That is, $\theta = \{c_0, \sigma_w^2, \mathbf{h}\}$ is the set of parameters to be estimated from a collection of consecutive observations $\mathbf{y} = \{y_0, y_1, \ldots, y_{N-1}\}$.

5.3 ▪ Blind Equalization Techniques

5.3.1 ▪ The EM-UKS Method

The above estimation problem can be addressed by the maximum likelihood approach. Since chaotic signals are generated by a dynamical system, EM is a natural choice for performing MLE iteratively [13]. EM is a two-stage process that involves an expectation step (E-Step) and a maximization step (M-Step), which is performed iteratively. We start from a random initial conditions and iterate until the difference between the likelihood values of two adjacent iterations is less than a user-defined threshold.

E-Step

In the EM algorithm, missing variables are introduced as a part of the estimation process. In certain situation this variable is introduced as an artifact to make the problem tractable. In some other situations, this missing variable will be a part of the estimation problem. In our problem, the state variables of the chaotic system are unobserved, and they naturally become the missing variables. We formulate an augmented state space model for the chaotic identification problem. The unknown parameter vector is θ, and the state vector $\mathbf{z}_n = [z_n, z_{n-1}, \ldots, z_{n-(p-1)}]$ follows the following dynamical equation:

$$\mathbf{z}_{n+1} = \begin{bmatrix} f(z_n) \\ 0 \end{bmatrix} + \begin{bmatrix} 0 & 0 \\ \mathbf{I}_{p-1} & 0 \end{bmatrix} \mathbf{z}_n \tag{5.4}$$

where $\mathbf{I}_{p-1}$ is an identity matrix of order $p-1$, and p is the order of the system. We assume that this information is available at the time of estimation. In the E-step we construct the complete statistics, $\mathscr{P}(\mathbf{Z}, \mathbf{y})$, with hidden variables $\mathbf{Z} = [\mathbf{z}_0, \ldots, \mathbf{z}_{N-1}]$ and observations $\mathbf{y} = [y_0, \ldots, y_{N-1}]$. The probability density function is then given by

$$\mathscr{P}(\mathbf{Z}, \mathbf{y}) = \mathscr{P}(\mathbf{z}_0) \prod_{n=1}^{N-1} \mathscr{P}(y_n | \mathbf{z}_n). \tag{5.5}$$

Due to the deterministic nature, $\mathscr{P}(\mathbf{z}_{n+1} | \mathbf{z}_n) = 1$. We have

$$\begin{aligned} \mathscr{P}(\mathbf{z}_1) &= \mathcal{N}(0, \sigma_c^2 \mathbf{I}), \\ \mathscr{P}(y_n | \mathbf{z}_n) &= \mathcal{N}(\mathbf{h}^T \mathbf{z}_n, \sigma_w^2). \end{aligned} \tag{5.6}$$

Once this is obtained, the next step is to get the expectation Q as

$$\begin{aligned} Q &= \mathbb{E}[\mathscr{L}(\mathbf{Z}, \mathbf{y}) | \mathbf{y}, \theta] \\ &= -\frac{1}{2} \ln(2\pi\sigma_c^2) - \frac{N}{2} \ln(2\pi\sigma_w^2) - \frac{1}{2\sigma_c^2} \mathbb{E}\left[\mathbf{x}_0^2 | \mathbf{y}, \theta\right] - \sum_{n=1}^{N-1} y_n^2 \\ &\quad -2 \sum_{n=1}^{N-1} y_n \mathbf{h}^T \mathbb{E}[\mathbf{z}_n | \mathbf{y}, \theta] + \sum_{n=1}^{N-1} \mathbf{h}^T \mathbb{E}\left[\mathbf{z}_n \mathbf{z}_n^T | \mathbf{y}, \theta\right] \mathbf{h}. \end{aligned} \tag{5.7}$$

In order to compute the above expectation, we need to find the individual expectations

$$\begin{aligned} \mathbf{x}_n^s &= \mathbb{E}[\mathbf{z}_n | \mathbf{y}, \theta], \\ \mathbf{P}_n^s &= \mathbb{E}[\mathbf{z}_n \mathbf{z}_n^T | \mathbf{y}, \theta]. \end{aligned} \tag{5.8}$$

A smoother can provide these individual expectation values. Since the sequence generated from a nonlinear system, the unscented Kalman smoother (UKS) is used here for its robustness [14]. The UKS is based on Rauch–Tung–Striebel smoothing [15]. The numerical instability of the covariance matrix is periodically checked and the algorithm is reinitialized accordingly.

M-Step

In the M-step, we use the results from the previous steps to perform maximization using the following steps:

$$\hat{\mathbf{h}} = \left[\sum_{n=0}^{N-1} P_n^s\right]^{-1} \sum_{n=1}^{N-1} y_n \mathbf{s}_n^s, \quad \hat{\sigma}_w^2 = \frac{1}{N} \sum_{n=0}^{N-1} (y_n - \hat{\mathbf{h}}^T \mathbf{s}_n^s), \quad \text{and } \hat{\sigma}_c = c_0^s. \tag{5.9}$$

These results will then be passed back for the next E-step iteration. The process is repeated until the algorithm converges. The schematic of the proposed method is shown in Figure 5.1.

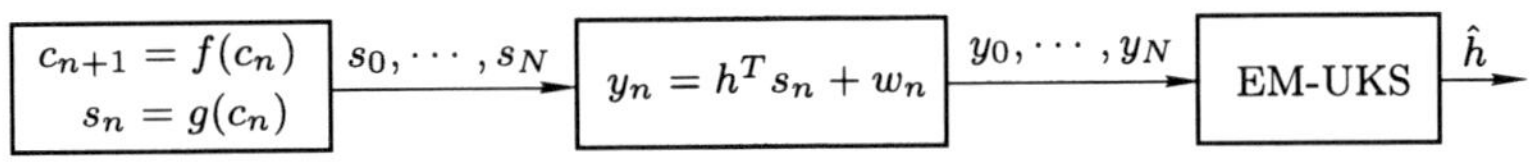

Figure 5.1. *The procedure of chaotic identification of FIR systems using EM-UKS.*

5.3.2 ▪ The Constant Modulus Algorithm

One of the most widely used blind equalization algorithms is the constant modulus algorithm (CMA) [16]. When the transmitted signal has constant modulus, we can use CMA for equalization. In fact, it has been shown that there are many other signals which do not have constant modulus but can also be equalized by the CMA. The CMA can be formulated as follows. The idea is very similar to the least mean square algorithm. We assume that the output of the system is passed through a linear filter such that

$$y_k = \mathbf{w}^H \mathbf{x}_k. \tag{5.10}$$

If we assume constant modulus for the input signal, one can formulate the estimation problem as

$$\min_{\mathbf{w}} J(\mathbf{w}), \text{ where } J(\mathbf{w}) = E\left[\left(|y_k|^2 - 1\right)^2\right]. \tag{5.11}$$

This can be solved by stochastic gradient method. That is,

$$\mathbf{w}_{k+1} = \mathbf{w} - \mu \nabla J(\mathbf{w}), \tag{5.12}$$

where $\nabla J(\mathbf{w})$ can be written as

$$\nabla J(\mathbf{w}) = 2E\left\{\left(|y_k|^2 - 1\right) y_k \mathbf{x}_k\right\}. \tag{5.13}$$

To avoid storing all the data for ensemble averaging, we can replace the expectation with its instantaneous value. The CMA can then be written as

$$\begin{aligned} y_k &= \mathbf{w}^H \mathbf{x}_k, \\ \mathbf{w}_{k+1} &= \mathbf{w} - \mu \mathbf{x}_k \left(|y_k|^2 - 1\right) y_k. \end{aligned} \tag{5.14}$$

5.3.3 ▪ Higher-Order Statistics-Based System Identification

The advantage of using higher-order statistics (HOS) for blind equalization is that HOS can retrieve both phase and amplitude [17]. We can see in the previous section that CMA has phase ambiguities and extra care needs to be taken. Another advantage of HOS-based algorithms is their abilities in estimating non–minimum phase systems [17]. Depending on the order of the cumulant, the algorithm can be roughly divided into two classes, namely, third-order cumulants (CUM3) and fourth-order cumulants (CUM4).

While the first-order cumulant of a stationary process is its mean, $C_{1,x} = \mathbb{E}[x_n]$, the higher-order cumulants can be defined as

$$\begin{aligned} C_{2,x}(k) &= \mathbb{E}\left[x_n^* x_{n+k}\right], \\ C_{3,x}(k,l) &= \mathbb{E}\left[x_n^* x_{n+k} x_{n+l}\right], \\ C_{4,x}(k,l,m) &= \mathbb{E}\left[x_n^* x_{n+k} x_{n+l} x_{n+m}^*\right] \\ &\quad - C_{2,x}(k) C_{2,x}(l-m) - C_{2,x}(l) C_{2,x}(k-m) \\ &\quad - M_{2,x}^*(m) M_{2,x}(k-l), \end{aligned} \tag{5.15}$$

where $M_{2,x}(m) = \mathbb{E}\left[x_n x_{n+m}\right]$ and equals $C_{2,x}(k)$ for a real-valued process. Cumulants have many interesting properties. For examples, if two signals x_n and w_n are statistically independent and $y_n = x_n + w_n$ we have

$$C_{4,y}(k,l,m) = C_{4,x}(k,l,m) + C_{4,w}(k,l,m). \tag{5.16}$$

This additive property can greatly simplify the cumulant analysis. We have $C_{k,w}([m_1, \ldots, m_k - 1]) = 0$ if $w_n$is Gaussian. Similarly, let $x_n = \sum_k h_k u_{n-k}$, where u_n is an i.i.d. input signal to a linear channel, $\mathbf{h} = \{h_0, h_2, \ldots, h_p\}$, we have the following relationship:

$$\begin{aligned} C_{2,x}(k) &= \gamma_{2,u} \sum_n h_n^* h_{n+k}, \\ C_{3,x}(k,l) &= \gamma_{3,u} \sum_n h_n^* h_{n+k} h_{n+l}, \\ C_{4,x}(k,l,m) &= \gamma_{4,u} \sum_n h_n^* h_{n+k} h_{n+l} h_{n+m}^*, \end{aligned} \tag{5.17}$$

where $\gamma_{k,u} = C_{k,u}(0)$. If u_n is non-Gaussian, one can retrieve the phase information from the cumulants and hence it can be used for the estimation of non-minimum phase signals. This is the property to be used for the parameter estimation using HOS. Before getting into the details of the algorithm, one needs to understand the practical aspects of the cumulant computation since we will have a finite number of samples. Typically, for system identification we will have N samples from which we have to compute the cumulants by the ensemble approximations given by

$$\begin{aligned} \hat{C}_{2,x}(k) &= \frac{1}{N_3} \sum_{n=N_1}^{N_2} x_n^* x_{n+k}, \\ \hat{M}_{x,y}(k) &= \frac{1}{N_3} \sum_{n=N_1}^{N_2} x_n^* y_{n+k}, \\ \hat{C}_{3,x}(k,l) &= \frac{1}{N_3} \sum_{n=N_1}^{N_2} x_n^* x_{n+k} x_{n+l}, \\ \hat{C}_{4,x}(k,l,m) &= \frac{1}{N_3} \sum_{n=N_1}^{N_2} x_n^* x_{n+k} x_{n+l} x_{n+m}^* \\ &\quad - \hat{C}_{2,x}(k) C_{2,x}(l-m) \\ &\quad - \hat{C}_{2,x}(l) C_{2,x}(k-m) - \hat{M}_{2,x}^*(m) M_{2,x}(k-l), \end{aligned} \tag{5.18}$$

where N_1 and N_2 are chosen such that the summation involves only x_n's with $n \in [0, N-1]$. Pan and Nikias [17] developed a method based on the cumulant relationship to estimate the system parameters when the input to the system is a non-Gaussian i.i.d. signal. The channel and cumulant relationship can then be written as

$$\sum_{m=-\infty}^{\infty} m\hat{h}_m \left[C_{3,y}(k-m,n)+C_{3,y}(k+m,l+m)\right] = kC_{3,y}(k,l). \tag{5.19}$$

Since the complex cepstrum is known to be exponentially bounded, the infinite summation can be reduced to a finite summation by setting $m=-q$ to $m=p$, where p and q are selected based on some prior knowledge. This leads to a set of linear equations that can be solved by computing all the required cumulants. Note that (5.19) does not solve for the case $m=0$. Since there is always a scalar ambiguity in this case, we let $\hat{h}_0 = 1$.

5.3.4 ▪ Minimum Nonlinear Prediction Error Method

So far, we discussed generic system identification techniques. Now, we will consider one of the previously developed chaotic identification methods, the minimum nonlinear prediction error technique [18]. The basic idea is that as the driving signal is chaotic, it can be modeled by a low-dimensional nonlinear map, and if the identification is performed properly, then the output of the equalizer should also follow that nonlinear map, at least in principle. Thus, identifying the system can be formulated as an optimization that minimizes the prediction error of the output signal. In this chapter, we extend the minimum prediction error method from an autoregressive model to a FIR system for system identification.

The inverse filtering technique estimates the system parameters by minimizing the cost function [18]

$$V_N(\theta) = \frac{1}{N}\sum_{n=1}^{N}\rho(n,\theta,u_n(\theta)), \tag{5.20}$$

where $u_n(\theta)$ is obtained through the inverse filtering of observation which is parameterized in θ. The cost function is defined as the mean square prediction error here, and (5.20) becomes

$$V_N(\theta) = \frac{1}{N}\sum_{n=1}^{N}(u_n(\theta)-x_n)^2. \tag{5.21}$$

Now the chaotic inverse filtering criterion can be written as

$$V_N(\theta) = \frac{1}{N-d+1}\sum_{n=1}^{N}\left[u_n(\theta)-g(\mathbf{u}_{n-1}(\theta))\right]^2, \tag{5.22}$$

where $\mathbf{u}_n = \left[u_{n-1}, u_{n-2}, \ldots, u_{n-d}\right]$ which is obtained by time-delay embedding. We use the inverse filter

$$u_n = y_n + \sum_{i=0}^{i=\infty}\hat{f}_i y_{n-i}. \tag{5.23}$$

In the absence of channel noise, it is equivalent to

$$u_n = \hat{F}(z^{-1})x_n, \tag{5.24}$$

where $\hat{F}(z^{-1}) = 1+\sum_{i=0}^{i=\infty}\hat{f}_i$. It is proven in [18] that the sufficient condition to minimize the prediction error is to have $\hat{F}(z^{-1}) = F(z^{-1})$. Here, $F(z^{-1}) = 1/B(z^{-1})$. In order to make the algorithm tractable, one could make the summation up to a constant l.

5.4 ▪ Performance Evaluation

In this section, simulations will be conducted to evaluate the system identification methods mentioned above. The logistic map given as follows will be used to generate chaotic sequences for system identification:

$$c_{n+1} = \gamma c_n(1 - c_n), \tag{5.25}$$

where $\gamma = 4$ is used in our simulations. In Figure 5.2, the phase space plot of the logistic map is depicted. For the symbolic sequence in (5.3), we set the threshold value $\mu = 0.5$.

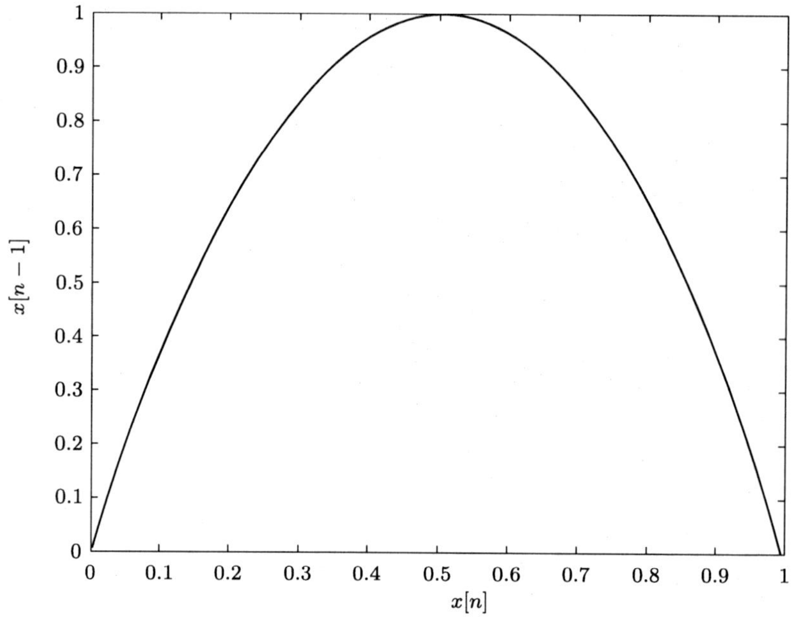

Figure 5.2. *The phase space plot of the trajectory of the logistic map.*

We define the estimation error in identifying a FIR system $\mathbf{h}$ as $\text{MSE}_{\mathbf{h}}$:

$$\text{MSE}_{\mathbf{h}} = \frac{1}{M} \sum_{i=1}^{M} \frac{||\mathbf{h} - \widehat{\mathbf{h}}||}{P}, \tag{5.26}$$

where M is the number of iterations to generate the SNR values. In the next section, we will derive the CRLB to assess the theoretical system identification performance. It is shown that under the asymptotic regime, the chaotic identification method can achieve the same performance as the nonblind system identification scheme.

5.4.1 ▪ The Cramér–Rao Lower Bound of the Chaotic Method

In this section, we derive the CRLB for the chaotic identification scheme. CRLB gives the minimum achievable estimator variance. By assuming independent measurements, the conditional probability $\mathbf{y}$ given θ for a generic perturbation signal $\mathbf{z}_n$ is given by

$$\mathscr{P}(\mathbf{y}|\theta) = \prod_{n=0}^{N-1} \frac{1}{\sqrt{2\pi\sigma_w^2}} \exp -\frac{(y_n - \mathbf{h}^T \mathbf{z}_n)^2}{2\sigma_w^2}, \tag{5.27}$$

and the corresponding log-likelihood function can be written as

$$\mathscr{L}(\mathbf{y}|\theta) = -\frac{N}{2}\ln(2\pi\sigma_w^2) - \frac{1}{2\sigma_w^2}\sum_{n=0}^{N-1}(y_n - \mathbf{h}^T \mathbf{z}_n)^2. \tag{5.28}$$

The CRLB of an estimator is given by the inverse of Fisher information matrix $\mathbf{J}(\theta)$ as

$$\mathrm{var}(\theta - \hat{\theta}) \geq [\mathbf{J}(\theta)]^{-1}, \tag{5.29}$$

where $\mathbf{J}(\theta)$ is given by

$$\mathbf{J}(\theta) = \mathbb{E}$$
$$\begin{bmatrix} \frac{\partial \mathscr{L}(\mathbf{y}|\theta)}{\partial \mathbf{h}}\left[\frac{\partial \mathscr{L}(\mathbf{y}|\theta)}{\partial \mathbf{h}}\right]^T & \frac{\partial \mathscr{L}(\mathbf{y}|\theta)}{\partial \mathbf{h}}\left[\frac{\partial \mathscr{L}(\mathbf{y}|\theta)}{\partial c_0}\right]^T & \frac{\partial^2 \mathscr{L}(\mathbf{y}|\theta)}{\partial \mathbf{h}}\left[\frac{\partial \mathscr{L}(\mathbf{y}|\theta)}{\partial \sigma_w}\right]^T \\ \frac{\partial \mathscr{L}(\mathbf{y}|\theta)}{\partial c_0}\left[\frac{\partial \mathscr{L}(\mathbf{y}|\theta)}{\partial \mathbf{h}}\right]^T & \frac{\partial \mathscr{L}(\mathbf{y}|\theta)}{\partial c_0}\left[\frac{\partial \mathscr{L}(\mathbf{y}|\theta)}{\partial c_0}\right]^T & \frac{\partial \mathscr{L}(\mathbf{y}|\theta)}{\partial c_0}\left[\frac{\partial \mathscr{L}(\mathbf{y}|\theta)}{\partial \sigma_w^2}\right]^T \\ \frac{\partial \mathscr{L}(\mathbf{y}|\theta)}{\partial \sigma_w}\left[\frac{\partial \mathscr{L}(\mathbf{y}|\theta)}{\partial \mathbf{h}}\right]^T & \frac{\partial \mathscr{L}(\mathbf{y}|\theta)}{\partial \sigma_w}\left[\frac{\partial \mathscr{L}(\mathbf{y}|\theta)}{\partial c_0}\right]^T & \frac{\partial \mathscr{L}(\mathbf{y}|\theta)}{\partial \sigma_w}\left[\frac{\partial \mathscr{L}(\mathbf{y}|\theta)}{\partial \sigma_w}\right]^T \end{bmatrix} \tag{5.30}$$

When we compute the first-order derivative of (5.28), we have

$$\begin{aligned} \frac{\partial \mathscr{L}(\mathbf{y}|\theta)}{\partial \mathbf{h}} &= \frac{1}{\sigma_w^2}\sum_{n=0}^{N-1}(y_n - \mathbf{h}^T \mathbf{z}_\mathrm{n})\mathbf{z}_n, \\ \frac{\partial \mathscr{L}(\mathbf{y}|\theta)}{\partial c_0} &= \frac{1}{\sigma_w^2}\sum_{n=0}^{N-1}(y_n - \mathbf{h}^T \mathbf{z}_\mathrm{n})\mathbf{h}^T \mathbf{z}'_n, \\ \frac{\partial \mathscr{L}(\mathbf{y}|\theta)}{\partial \sigma_w} &= -\frac{N}{\sigma_w} + \frac{1}{\sigma_w^3}\sum_{n=0}^{N-1}(y_n - \mathbf{h}^T \mathbf{z}_\mathrm{n})^2, \end{aligned} \tag{5.31}$$

and the second derivatives are given by

$$\begin{aligned} \frac{\partial \mathscr{L}(\mathbf{y}|\theta)}{\partial \mathbf{h}}\left[\frac{\partial \mathscr{L}(\mathbf{y}|\theta)}{\partial \mathbf{h}}\right]^T &= -\frac{1}{2\sigma_w^2}\sum_{n=0}^{N-1}\mathbf{z}_n\mathbf{z}_n^T, \\ \frac{\partial \mathscr{L}(\mathbf{y}|\theta)}{\partial c_0}\left[\frac{\partial \mathscr{L}(\mathbf{y}|\theta)}{\partial \mathbf{h}}\right]^T &= \frac{1}{\sigma_w^2}\sum_{n=0}^{N-1}(w_n\mathbf{z}_{n'} - \mathbf{z}_n[\mathbf{z}_{n'}]^T\mathbf{h}), \\ \frac{\partial \mathscr{L}(\mathbf{y}|\theta)}{\partial \sigma_w}\left[\frac{\partial \mathscr{L}(\mathbf{y}|\theta)}{\partial \mathbf{h}}\right]^T &= -\frac{2}{\sigma_w^2}\sum_{n=0}^{N-1}w_n\mathbf{z}_n, \end{aligned}$$

$$\frac{\partial^2 \mathscr{L}(\mathbf{y}|\theta)}{\partial \mathbf{h}\partial c_0} = \frac{1}{\sigma_w^2}\sum_{n=0}^{N-1}\left(w_n\mathbf{z}_{n'} - \mathbf{h}^T\mathbf{z}_n\mathbf{z}_{n'}\right),$$

$$\frac{\partial^2 \mathscr{L}(\mathbf{y}|\theta)}{\partial^2 c_0} = \frac{1}{\sigma_w^2} \sum_{n=0}^{N-1} \left(w_n \mathbf{h}^T \mathbf{z}_{n''} - (\mathbf{h}^T \mathbf{z}_{n'})^2 \right),$$

$$\frac{\partial^2 \mathscr{L}(\mathbf{y}|\theta)}{\partial \sigma_w \partial c_0} = -\frac{2}{\sigma_w^3} \sum_{n=0}^{N-1} w_n \mathbf{h}^T \mathbf{z}_{n'},$$

$$\frac{\partial^2 \mathscr{L}(\mathbf{y}|\theta)}{\partial \mathbf{h} \partial \sigma_w} = -\frac{2}{\sigma_w^3} \sum_{n=0}^{N-1} w_n \mathbf{z}_n,$$

$$\frac{\partial^2 \mathscr{L}(\mathbf{y}|\theta)}{\partial c_0 \partial \sigma_w} = -\frac{2}{\sigma_w^3} \sum_{n=0}^{N-1} w_n \mathbf{h}^T \mathbf{z}'_n, \tag{5.32}$$

$$\frac{\partial^2 \mathscr{L}(\mathbf{y}|\theta)}{\partial^2 \sigma_w} = \frac{N}{\sigma_w^2} - \frac{3}{\sigma_w^4} \sum_{n=0}^{N-1} w_n^2,$$

where $\mathbf{z}_{n'}$ and $\mathbf{z}_{n''}$ are the first and second derivative of $\mathbf{z}_n$ with respect to c_0. Since the w_n is a zero mean Gaussian process, all the terms that contain w_n become zero and terms that contain w_n^2 become σ_w^2 when we take the expectation. $\mathbf{J}(\theta)$ can then be expressed as

$$\frac{1}{\sigma_w^2} \begin{pmatrix} \mathbb{E}\left[\sum_{n=0}^{N-1} \mathbf{z}_n \mathbf{z}_n^T\right] & \mathbb{E}\left[\sum_{n=0}^{N-1} \mathbf{z}_n [\mathbf{z}_{n'}]^T \mathbf{h}\right] & 0 \\ \mathbb{E}\left[\sum_{n=0}^{N-1} \mathbf{h}^T \mathbf{z}_n \mathbf{z}_{n'}\right] & \mathbb{E}\left[\sum_{n=0}^{N-1} (\mathbf{h}^T \mathbf{z}_{n'})^2\right] & 0 \\ 0 & 0 & 2N \end{pmatrix}. \tag{5.33}$$

If we represent the above matrix with $\mathbf{J}_{11} = \mathbb{E}\left[\sum_{n=0}^{N-1} \mathbf{z}_n \mathbf{z}_n^T\right]$, $\mathbf{a} = \mathbb{E}\left[\sum_{n=0}^{N-1} \mathbf{z}_n [\mathbf{z}_{n'}]^T \mathbf{h}\right]$, and $J_{22} = \mathbb{E}\left[\sum_{n=0}^{N-1} (\mathbf{h}^T \mathbf{z}_{n'})^2\right]$, then the CRLB can be written as

$$\mathrm{var}(\theta - \hat{\theta}) \geq \sigma_w^2 \begin{pmatrix} \mathbf{J}_{11} & \mathbf{a} & 0 \\ \mathbf{a}^T & J_{22} & 0 \\ 0 & 0 & 2N \end{pmatrix}^{-1} = \sigma_w^2 \begin{bmatrix} \begin{pmatrix} \mathbf{J}_{11} & \mathbf{a} \\ \mathbf{a}^T & j_{22} \end{pmatrix}^{-1} & 0 \\ 0 & \frac{1}{2N} \end{bmatrix}. \tag{5.34}$$

Using the block matrix inverse identity we can rewrite the above inequality as

$$\mathrm{var}(\theta - \hat{\theta}) \geq \begin{pmatrix} \Delta^{-1} & -\Delta^{-1}\mathbf{a} j_{22}^{-1} & 0 \\ -j_{22}^{-1}\mathbf{a}^T \Delta^{-1} & j_{22}^{-1} + j_{22}^{-1}\mathbf{a}^T \Delta^{-1} \mathbf{a} j_{22}^{-1} & 0 \\ 0 & 0 & \frac{1}{2N} \end{pmatrix}. \tag{5.35}$$

where $\Delta = \mathbf{J}_{11} - \mathbf{a} j_{22}^{-1} \mathbf{a}^T$. We are interested in the error variance $\mathbb{E}[\Delta\mathbf{h}^T \Delta\mathbf{h}]$. Here $\Delta\mathbf{h}$ is the estimate where

$$\mathrm{Tr}[\Delta\mathbf{h}^T \Delta\mathbf{h}] \geq \mathrm{Tr}(\Delta^{-1}) = \mathrm{Tr}\left([\mathbf{J}_{11} - \mathbf{a} j_{22}^{-1} \mathbf{a}^T]^{-1}\right). \tag{5.36}$$

5.4.2 ▪ The Cramér–Rao Lower Bound of the Nonblind Identification Method

For the nonblind case, the parameters need to be estimated are $\mathbf{h}$ and σ_w. Calculating the first and second derivatives and forming the FIM, the CRLB becomes

$$\mathrm{var}(\hat{\theta}) \geq \sigma_w^2 \left\{ \mathbb{E} \begin{bmatrix} \sum_{n=0}^{N-1} \mathbf{x}_n \mathbf{x}_n^T & 0 \\ 0 & 2N \end{bmatrix} \right\}^{-1}. \tag{5.37}$$

The desired autocorrelation function for the excitation signal is Dirac delta function $\delta(t)$. This means that $\mathbf{E}\left[\sum_{n=0}^{N-1} \mathbf{x}_n \mathbf{x}_n^T\right] = N\mathbf{I}_{p\times p}\sigma_u^2$, where σ_u^2 is the power of the input sequence. When a white Gaussian noise sequence is used as the excitation signal, we have $\mathbb{E}[\Delta\mathbf{h}^T\Delta\mathbf{h}] = \frac{p\sigma_w^2}{N\sigma_u^2}$.

Theorem: *For any FIR system identification, the asymptotic CRLB of the nonblind system and chaotic sequence-based blind identification proposed are equivalent.*

Proof: Proof of this theorem can be obtained in two steps. In the first step we show that $\mathrm{Tr}\left([\mathbf{J}_{11} - \mathbf{a} j_{22}^{-1}{}' \mathbf{a}^T]^{-1}\right)$ tends to $\mathrm{Tr}[\mathbf{J}_{11}]^{-1}$ for a large N. Let us consider the value of j_{22}, that is,

$$j_{22} = \sum_{n=0}^{N-1} \mathbb{E}\left[\frac{1}{\sigma_w^2}\sum_{n=0}^{N-1}(\mathbf{h}^T \mathbf{z}_{n'})^2\right]$$

$$= \sum_{n=0}^{N-1} \mathbf{h}^T \mathbb{E} \begin{bmatrix} \left|\frac{\partial z_n}{\partial c_0}\right|^2 & \frac{\partial z_n}{\partial z_{n-1}}\left|\frac{\partial z_{n-1}}{\partial c_0}\right|^2 & \cdots \\ \frac{\partial z_n}{\partial z_{n-1}}\left|\frac{\partial z_{n-1}}{\partial c_0}\right|^2 & \left|\frac{\partial z_{n-1}}{\partial c_0}\right|^2 & \cdots \\ \cdots & \cdots & \left|\frac{\partial z_{n-p+1}}{\partial c_0}\right|^2 \end{bmatrix} \mathbf{h} \tag{5.38}$$

$$= \sum_{n=0}^{N-1} \mathbb{E}\left[\frac{1}{\sigma_w^2}\sum_{n=1}^{N}(\mathbf{h}^T \mathbf{z}_{n'})^2\right]$$

$$= \sum_{n=0}^{N-1} \mathbf{h}^T \mathbb{E} \begin{bmatrix} \left|\frac{\partial z_n}{\partial c_0}\right|^2 & \frac{\partial z_n}{\partial z_{n-1}}\left|\frac{\partial z_{n-1}}{\partial c_0}\right|^2 & \cdots \\ \frac{\partial z_n}{\partial z_{n-1}}\left|\frac{\partial z_{n-1}}{\partial c_0}\right|^2 & \left|\frac{\partial z_{n-1}}{\partial c_0}\right|^2 & \cdots \\ \cdots & \cdots & \left|\frac{\partial z_{n-p+1}}{\partial c_0}\right|^2 \end{bmatrix} \mathbf{h}.$$

By taking the trace on both sides, we get a simplified expression for j_{22} where $h = \mathbf{h}^T\mathbf{h}$:

$$\begin{aligned}
j_{22} &= \mathrm{Tr}(j_{22}) \\
&= \sum_{n=0}^{N-1} \mathrm{Tr}\left(\mathbf{h}^T \mathbb{E}\begin{bmatrix} \left|\frac{\partial z_n}{\partial c_0}\right|^2 & \frac{\partial z_n}{\partial z_{n-1}}\left|\frac{\partial z_{n-1}}{\partial c_0}\right|^2 & \cdots \\ \frac{\partial z_n}{\partial z_{n-1}}\left|\frac{\partial z_{n-1}}{\partial c_0}\right|^2 & \left|\frac{\partial z_{n-1}}{\partial c_0}\right|^2 & \cdots \\ \cdots & \cdots & \left|\frac{\partial z_{n-p+1}}{\partial c_0}\right|^2 \end{bmatrix} \mathbf{h} \right) \\
&= h \sum_{n=0}^{N-1} \mathrm{Tr}\left(\mathbb{E}\begin{bmatrix} \left|\frac{\partial z_n}{\partial c_0}\right|^2 & \frac{\partial z_n}{\partial z_{n-1}}\left|\frac{\partial z_{n-1}}{\partial c_0}\right|^2 & \cdots \\ \frac{\partial z_n}{\partial z_{n-1}}\left|\frac{\partial z_{n-1}}{\partial c_0}\right|^2 & \left|\frac{\partial z_{n-1}}{\partial c_0}\right|^2 & \cdots \\ \cdots & \cdots & \left|\frac{\partial z_{n-p+1}}{\partial c_0}\right|^2 \end{bmatrix} \right) \\
&= h \sum_{n=0}^{N-1} \mathbb{E}\left[\left(\left|\frac{\partial z_n}{\partial c_0}\right|^2 + \ldots + \left|\frac{\partial z_{n-(p+1)}}{\partial c_0}\right|^2 \right) \right].
\end{aligned} \tag{5.39}$$

From the chaos theory, $\mathbb{E}\left[\left|\frac{\partial z_{n-i}}{\partial c_0}\right|^2\right] = \exp(2(n-i)\Lambda(n-1))$, where $\Lambda(n-1)$ is the local Lyapunov exponent of the chaotic map, we have

$$\lim_{N\to\infty} \sum_{n=0}^{N-1} \mathbb{E}\left[\left|\frac{\partial z_{n-i}}{\partial c_0}\right|^2\right] = \sum_{n=0}^{N-1} \exp\left(2(n-i)\Lambda(n-1)\right) = \infty. \tag{5.40}$$

In other words, $\lim_{N\to\infty} j_{22} = \infty$. This means $\mathbb{E}[\Delta\mathbf{h}^T\Delta\mathbf{h}] = \mathbf{J}_{11}^{-1}$. Now, it is sufficient to prove that $\mathrm{Tr}(\mathbf{J}_{11}^{-1})$ is equal to $\frac{\sigma_w^2}{N\sigma_c^3}$. To do that, we write

$$\begin{aligned}
\mathbf{J}_{11} &= \mathbb{E}\left[\sum_{n=0}^{N-1} \mathbf{c}_n \mathbf{c}_n^T\right] \\
&= \sum_{n=1}^{N} \begin{bmatrix} \mathbb{E}[c_n c_n] & \mathbb{E}[c_n c_{n-1}] & \cdots \\ \mathbb{E}[c_n c_{n-1}] & \mathbb{E}[c_{n-1} c_{n-1}] & \cdots \\ \cdots & \cdots & \mathbb{E}[c_{n-(p+1)} c_{n-(p+1)}] \end{bmatrix}. \\
&\approx \sigma_c^2 \mathbf{I}_{p\times p}
\end{aligned} \tag{5.41}$$

Since chaotic sequences have impulselike autocorrelation function

$$\mathbb{E}\left[\sum_{n=0}^{N-1} \mathbf{c}_{n-i}\mathbf{c}_{n-j}^T\right] \approx \begin{cases} \sigma_c^2 & \text{for } i = j, \\ 0 & \text{for } i \neq j, \end{cases} \tag{5.42}$$

where σ_c^2 is the power of chaotic sequence, it can now be concluded that when $\sigma_u^2 = \sigma_c^2$, the CRLB of the blind chaotic identification scheme and the nonblind identification are equivalent.

5.4.3 ▪ Computer Simulations

We now consider the system identification problem in a general setting with FIR coefficients $\mathbf{h} = [1, \ 0.45, \ -0.2]$. Frequency response of the channel is shown in Figure 5.3. MSE_h is calculated according to (5.26) with $N = 256$. The result is plotted in Figure 5.4. For comparison, we also plot the nonblind identification schemes based on white Gaussian noise and pseudorandom sequence. These results are compared with the theoretical

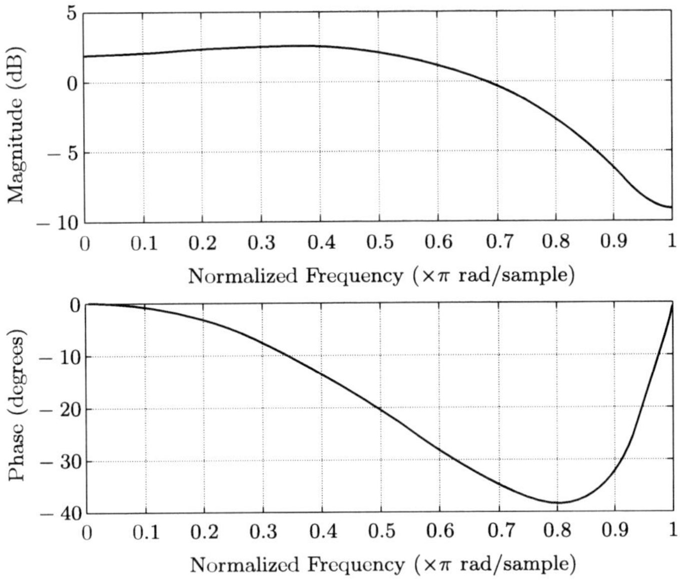

Figure 5.3. *Frequency response of the channel used for identification.*

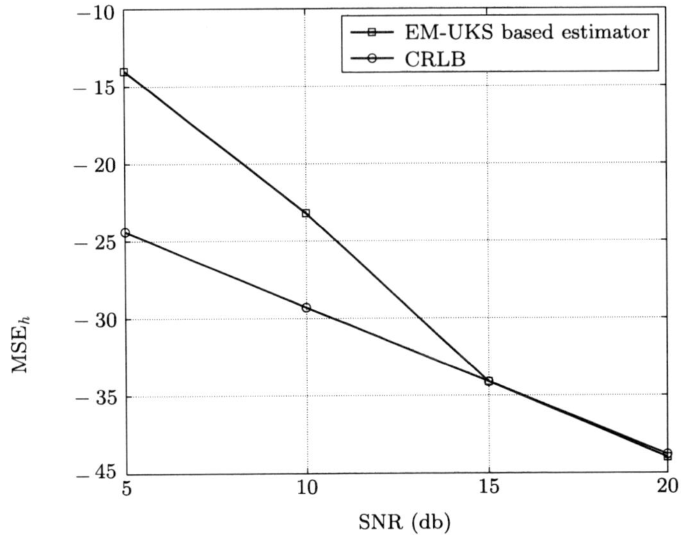

Figure 5.4. *MSE performance of EM-UKS–based estimation scheme for chaotic numeric sequence.*

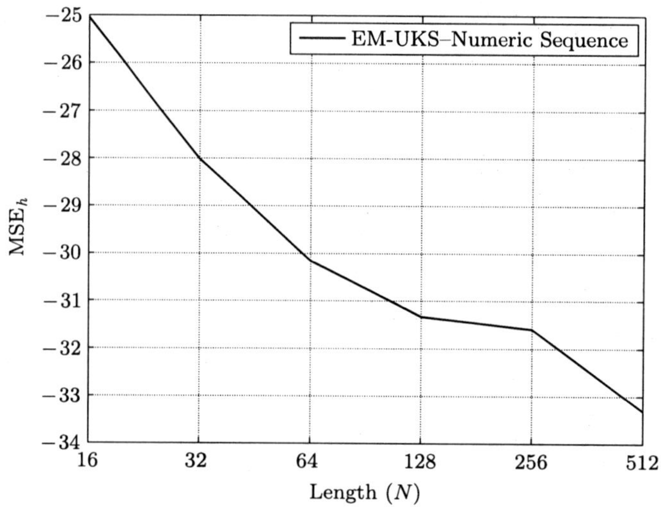

Figure 5.5. *MSE performance versus input sequence length.*

CRLB. The chaos-based estimation method is shown to follow the CRLB closely when the SNR is relatively high.

To study the effect of sequence length on the identification, we compute MSE_h using different sequence lengths, and the results are plotted in Figure 5.5. It can be seen that the accuracy depends on the sequence length, and the computational complexity also increases with the sequence length. At 15dB, we compute the MSE estimate for different lengths starting from 16. It can be seen that, even for fairly short sequence length, the proposed system has a good system identification performance with an estimation error of −25dB.

As an iterative algorithm, computational power requirement for the proposed EM method is relatively high. It is desirable to achieve convergence with minimal iterations. To understand the noise effect on the EM convergence, we perform a number of EM-UKS–based parameter estimations at two SNR values. The histograms of number of iterations taken for convergence are plotted in Figure 5.6. It can be seen that there is a strong dependency on the convergence of EM-UKS and the noise.

Now, we compare the performance of the proposed method with the conventional blind system identification techniques mentioned above; in particular, CMA and the higher-order statistics method using third-order cumulant (CUM3) and fourth-order cumulant (CUM4) are used in our simulations. From Figures 5.7 and 5.8, among all three methods considered, only the CUM4 algorithm is capable of estimating channel parameters effectively, but only at high SNRs.

Figure 5.9 shows the performance comparison of proposed method against the minimum nonlinear prediction error method. For all SNRs, the proposed method has a better MSE performance. For SNR higher than 30dB, the performance of the proposed method flattens out a bit because we did not use the same threshold of 0.001 for the EM existing criteria, which are relatively high. This threshold is selected to speed up the simulation. As a note, we observed that when the threshold is further reduced, a better performance can be achieved.

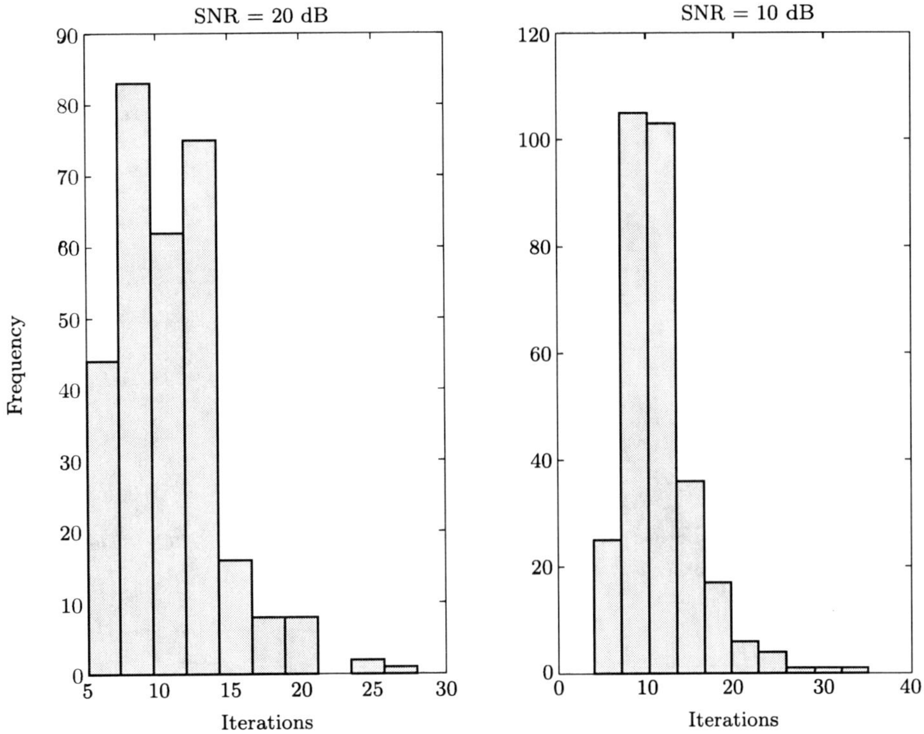

Figure 5.6. *Histogram of number of iterations taken for convergence:* (a) 20*dB and* (b) 10*dB.*

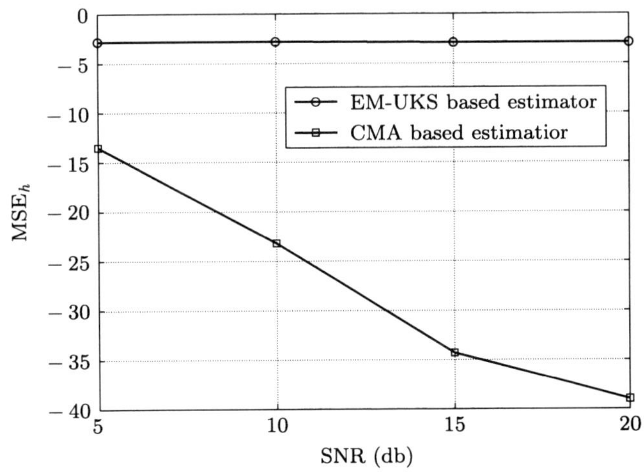

Figure 5.7. *Comparison of MSE performance of the proposed method and the CMA equalization scheme.*

As many applications in system identification involve binary sequences or sequences taking on finite states, we now consider the use of chaotic symbolic sequence instead of chaotic numerical sequence [19]. Symbolic sequences are the coarse-grain representation of the numerical sequences. Such sequences can be generated from the numerical

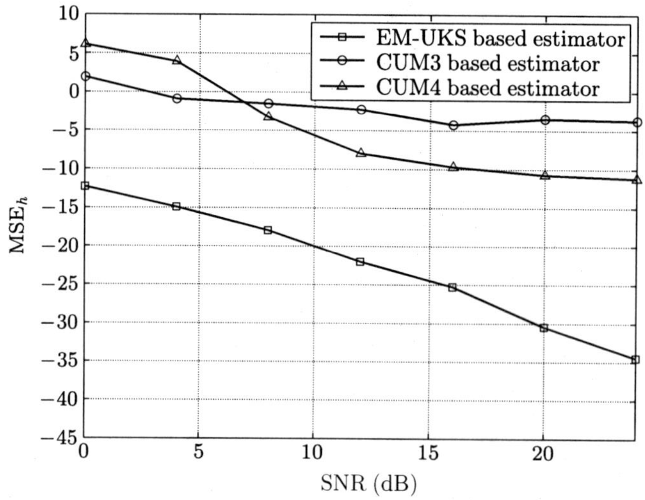

Figure 5.8. *Comparison of MSE performance of the proposed identification method with higher-order statistics-based schemes.*

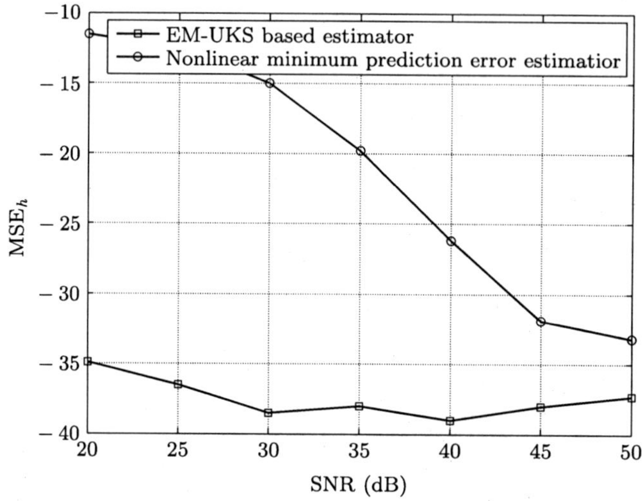

Figure 5.9. *Comparison of MSE performance of the proposed identification method with the minimum nonlinear prediction error method.*

sequences using appropriate phase space partitioning. Chaotic symbolic sequences are used in many applications ranging from spread-spectrum communications to cryptography [20, 21]. It is shown that the chaotic dynamics can act as a source of information and thus these sequences can be used as an alternative to the pseudorandom noise signal [22, 23].

Figure 5.10 shows the MSE performance for symbolic sequence with 128 bits length. For all SNRs, the MSE closely follows the CRLB. Figure 5.11 is plotted the MSE performance versus sequence length. As in the case of numerical sequences, we observe a steady drop in MSE when the length increases. However, the drop is steeper when symbolic sequences are used.

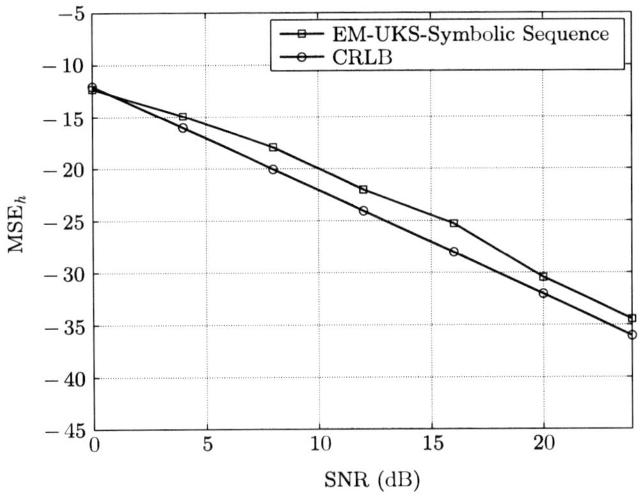

Figure 5.10. *MSE performance of the proposed identification method using symbolic sequences.*

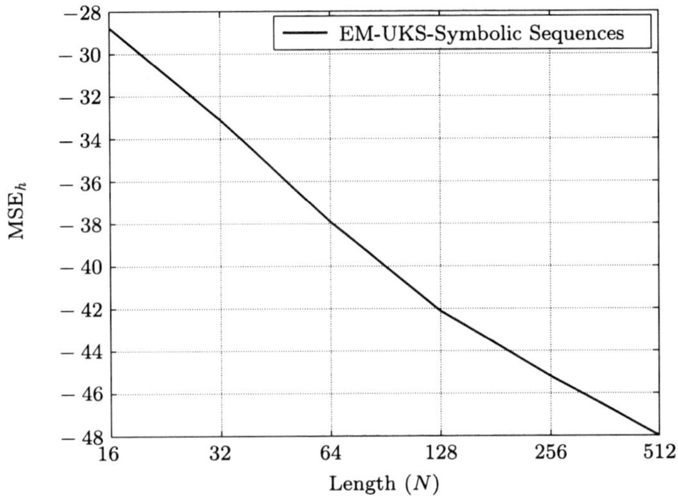

Figure 5.11. *MSE of the proposed identification method versus sequence length for symbolic sequences.*

5.4.4 ▪ Comparison of Chaotic Numerical Sequences and Symbolic Sequence

We study the average cross correlation function and crest factor for arbitrary realization of both numerical and symbolic sequences. The correlation function of the numerical chaotic sequence ($\mathbf{c} = [c_1, \ldots, c_N]$) and symbolic sequence ($\mathbf{s} = [g(c_1), \ldots, g(c_N)]$) is shown in Figure 5.12. This figure clearly shows that for finite chaotic sequences, the correlation function of the chaotic symbolic sequence has a better peak to valley ratio compared to its numerical counterpart. In other words, the symbolic sequences can have even better frequency spectra for system identification compared to the numerical sequences. The crest factor for symbolic sequence with two alphabets (binary sequence) is 1. From

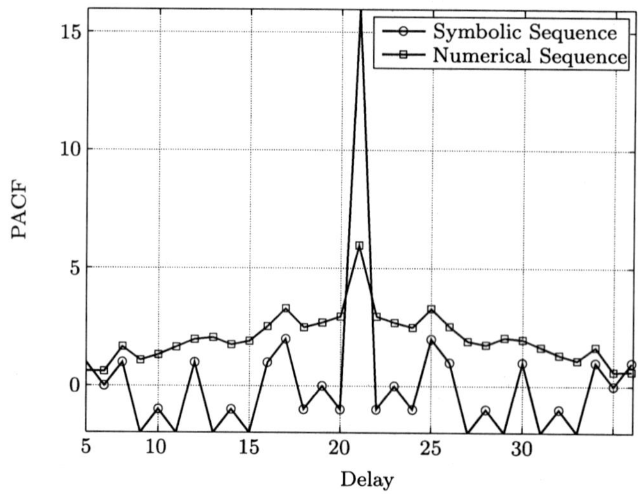

Figure 5.12. *Autocorrelation function of chaotic symbolic sequence and numerical sequence.*

the simulations, we found that the crest factor for the chaotic numerical sequence is 1.904. Clearly, the chaotic symbolic sequence transfers more energy to the system, which results in a better estimation performance.

It is noted that not all chaotic initial conditions will give symbolic sequences that possess the persistent excitation property. Selections of initial conditions that lead to code which will excite the system in all the frequencies are important.

The initial condition selection method used in our study is outline here. First, many initial conditions are generated randomly, and the corresponding symbolic sequences are constructed. We form the matrix **T**, which is a Toeplitz matrix with column s and first row $[s(1)\ \ 0]$. The condition number of $\mathbf{T}^T\mathbf{T}$ is calculated. We select initial conditions that result in low values of condition number.

5.5 ▪ Application to Noncoherent Ranging

Ranging is the process of measuring the distance between the transmitter and an object. Pseudo-noise sequences with the correlator as the receiver have been extensively used for ranging applications [11, 24]. Spread-spectrum–based ranging has attracted attention because the effective pulse energy is spread to a larger bandwidth, resulting in an accurate range estimate. However, such a system requires complex synchronization circuits. Especially, the incoming pulse and the receiver should be matched, at least within half a pulse width. Also, the local oscillator should have synchronized clocks with the transmitter. Poor synchronization can result in very inaccurate range estimation. Ranging in multipath is even more difficult, and a noncoherent mode for the estimation of delay is desired. We apply the proposed chaotic identification method to noncoherent ranging here. We consider a complete asynchronous ranging where there is no need for a tight synchronization of transmitter and receiver.

The block diagram of the ranging is shown in Figure 5.13. There are two units A and B, two sensor nodes which can transmit and receive radio frequency signals. The locations

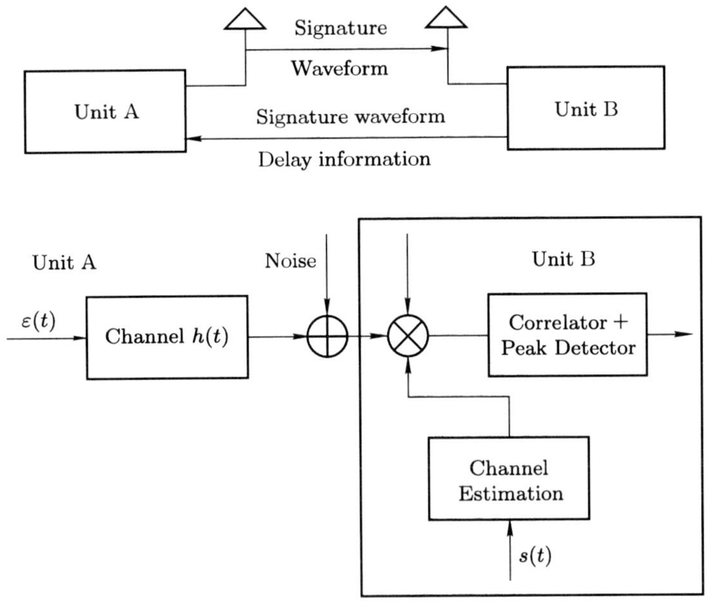

Figure 5.13. *Block diagram of a two-way range system:* (a) *signaling between the units and* (b) *waveform decoding.*

of these units are such that there is no direct line of sight communication. The objective of our analysis is to find the distance between the two units by sending a signature radio frequency waveform which is known by both units using a two way communication. Unit A sends the signature waveform and starts a counter. Unit B decodes the waveform and then sends it back to unit A along with the delay information. Once the signal is decoded at unit A correctly, the timer of unit A is stopped. By subtracting the time required for signal processing at unit A and unit B from the total timer duration, one can compute the round trip delay and then compute the distance between units A and B.

Let the chaotic symbolic sequence used for ranging is s_n. We assume a rectangular pulse shaping with $p(t)$, where

$$p(t)=\begin{cases} 1 & \text{if } t \le T_c, \\ 0 & \text{otherwise.} \end{cases} \tag{5.43}$$

In the above equation T_c is the pulse width. The transmitted signal after modulation becomes

$$r(t)=\sum_{k=1}^{n} s_n p(t-kT_c)\cos(\omega t), \tag{5.44}$$

where ω is the carrier frequency. The signal that travels through the multipath channel can be represented as

$$y(t)=h(t)*r(t), \tag{5.45}$$

and the reflectivity of the material is absorbed within $h(t)$. By adding the channel noise, we have

$$z(t)=y(t-\tau)+n(t). \tag{5.46}$$

Here, τ denotes the time taken for the signature waveform to travel from unit A to unit B. $z(t)$ can be demodulated using the local oscillator which may have different operating frequencies and phases compared to the modulating waveform. This demodulated signal is fed to the waveform decoder to decode the signature waveform. Before it is sent to the decoder, a channel estimation scheme is used to determine the channel effect between units A and B. This information is used for the decoding of the received signal by the peak detection unit in Figure 5.13. The peak of the correlation gives the estimate of the delay.

In the simulation, we compare the performance of the chaos-based range estimation using symbolic sequences against the higher-order statistics based range estimation scheme. We use a sequence length of 256, and each symbol is mapped to a rectangular wave pulse. Each of these pulses has four samples. Since the pulse width can affect the range resolution, we show the results in terms of the number of samples that the cross correlation peak is differ from the actual peak (for simplicity, we assume that it is at zero delay). We run these simulations for 1000 times for all SNR values. The average number of samples offset against the SNR values is plotted in Figure 5.14. At low SNR, CUM4 estimation technique has poor channel estimation performance. This leads to poor range estimates. As we expect, when the SNR increases, the range estimations of both estimators improve. While the figure shows the average results, we found that the number of estimation error can increase to 1015 samples at low SNR for the CUM4 estimation method, while for the chaos-based ranging scheme it is just 153 samples. From these analyses, we observe that the chaos approach is suitable for range estimation.

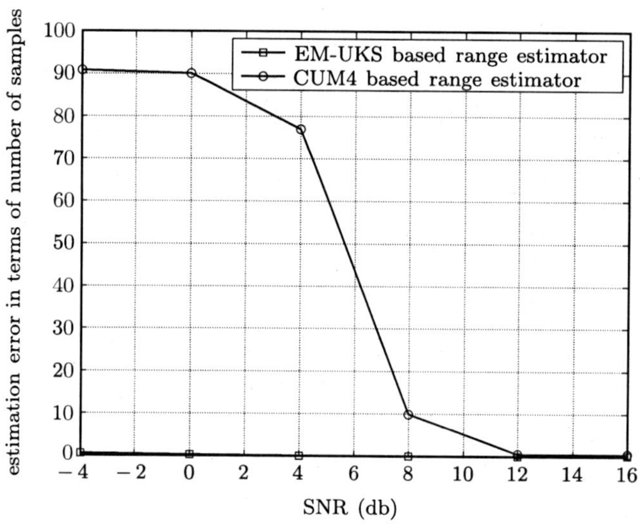

Figure 5.14. *Comparison of the chaos-based ranging method with the higher-order statistic-based method for ranging.*

5.6 ▪ Conclusions

One of the most challenging tasks in system identification is the input signal design. In this chapter, we used a maximum likelihood approach for system identification when the input to the system is generated from a chaotic dynamic system. Combining EM with UKS we formed an EM-UKS estimator to perform blind identification of an FIR system. We showed here that the asymptotic CRLB of the proposed chaos method is close to the nonblind system identification schemes. By comparing to the conventional blind identification schemes based on higher-order statistics, the proposed chaos identification approach was shown to have superior identification performance. We then applied the proposed method to systems driven by chaotic symbolic sequences and the good identification performance was also observed. As a practical application, we extended this chaos identification scheme for noncoherent ranging. From our analysis, we conclude that chaotic signals in either numerical or symbolic format can act as an ideal candidate for system identification and result in good identification performance.

References

[1] K. Abed-Meraim, W. Qiu, and Y. Hua, "Blind system identification," *Proc. IEEE*, 85, pp. 1310–1322, 1997.

[2] C. Chi, C. Feng, C. Chen, and C. Chen, *Blind Equalization and System Identification: Batch Processing Algorithms, Performance and Applications*, Springer, 2006.

[3] H. Papadopoulos and G. Wornell, "Optimal detection of a class of chaotic signals," *IEEE Intl. Conf. Acoust., Speech, Signal Processing*, vol. 3, pp. 117–120, 1993.

[4] N. Xie and H. Leung, "A nonlinear prediction approach for system identification using chaos symbolic dynamic," *IEEE Intl. Conf. Systems, Man and Cybernetics*, vol. 2, pp. 1365–1370, 2003.

[5] N. Xie and H. Leung, "Blind identification of autoregressive system using chaos," *IEEE Trans. Circuits Syst. I*, 52, pp. 1953–1964, 2005.

[6] L. M. Pecora and T. L. Carroll, "Synchronization in chaotic systems," *Phys. Rev. E*, 64, pp. 821–824, 1990.

[7] A. P. Dempster, N. M. Laird, and D. B. Rubin, "Maximum likelihood from incomplete data via the EM algorithm," *J. Roy. Statist. Soc. Ser. B*, 39, pp. 1–38, 1977.

[8] V. Venkatasubramanian and H. Leung, "Chaos based semi blind identification of nonlinear systems," *IEEE Intl. Workshop on Nonlinear Signal and Image Processing*, 2005.

[9] Y. Cai and T. Wang, "Analysis and design of binary pseudorandom sequences based on the Magma package," *International Conference on Electrical and Control Engineering (ICECE)*, 2011.

[10] M. K. Reynolds, M. J. Reinhart, R. S. Bokulic, and S. H. Bryant, "A two-way noncoherent ranging technique for deep space missions," *IEEE Aerospace Conference Proceedings*, 2002.

[11] H. A. Barker, D. E. Rivera, A. H. Tan, and K. R. Godfrey, "Perturbation signal design," 14th *IFAC Symp. System Identification*, 2006.

[12] L. Ljung, *System Identification—Theory for the User*, 2nd ed., PTR Prentice Hall, Upper Saddle River, NJ, 1999.

[13] R. H. Shumway and D. S. Stoffer, "An approach to time series smoothing and forecasting using the EM algorithm," *J. Time Series Anal.*, 3, pp. 253–264, 1982.

[14] E. A. Wan, R. van der Merwe, and A. T. Nelson, "Dual estimation and the unscented transformation," *Neural Information Processing Systems*, 12, pp. 666–672, 2000.

[15] S. Sarkka, "Unscented Rauch–Tung–Striebel smoother," *IEEE Trans. Automat. Control*, 53, pp. 845–849, 2008.

[16] R. Johnson, Jr., P. Schniter, T. Endres, J. Behm, D. Brown, and R. Casas, "Blindequalization using the constant modulus criterion: A review," *Proc. IEEE*, 86, pp. 1927–1950, 1998.

[17] R. Pan and C. L. Nikias, "The complex cepstrum of higher-order cumulants and non-minimum phase system identification," *IEEE Trans. Acoustics, Speech and Signal Processing*, 36, pp. 186–205, 1988.

[18] Z. Zhu and H. Leung, "Identification of linear systems driven by chaotic signals using nonlinear prediction," *IEEE Trans. Circuits and Syst. I*, 49, pp. 170–180, 2002.

[19] H. Bailin, *Elementary Symbolic Dynamics and Chaos in Dissipative Systems*, World Scientific, Singapore, 1989.

[20] S. Chen and H. Leung, "Chaotic watermarking for video authentication in surveillance applications," *IEEE Trans. Circuits Systems Video Technology*, 18, pp. 704–709, 2008.

[21] M. S. Baptista, "Cryptography with chaos," *Phys. Lett. A*, 240, pp. 50–54, 1998.

[22] R. Rovatti, G. Mazzini, and G. Setti, "On the ultimate limits of chaos-based asynchronous DS-CDMA-I: basic definitions and results," *IEEE Trans. Circuits Syst. I*, 51, pp. 1336–1347, 2004.

[23] R. Rovatti, G. Mazzini, and G. Setti, "On the ultimate limits of chaos-based asynchronous DS-CDMA-II: Analytical results and asymptotics," *IEEE Trans. Circuits Syst. I*, 51, pp. 1348–1364, 2004.

[24] D. Wang, R. Kannan, L. Wei, and B. Tay, "Time of flight based two way ranging for real time locating systems," *IEEE Conference on Robotics Automation and Mechatronics*, pp. 199–205, 2010.

Chapter 6

Characterization and Optimization of a Chaotic LADAR System for High Resolution Range Determination

Berenice Verdin and Benjamin C. Flores

Abstract: The purpose of this chapter is to study the system parameters and signal processing schemes required to build a high resolution chaotic LADAR (Laser Detection and Ranging). Specifically, the LADAR is based on a semiconductor laser that generates a wideband chaotic signal via optical feedback. The behavior of the laser output is modeled by changing the control parameters of the delay differential equations that dictate the behavior of the laser output. Depending on the feedback level and the external cavity length the output of the laser, a route to chaotic behavior is found. Experiments show that the power spectral density and autocorrelation of the electric field of the laser output can be controlled to achieve maximum range resolution. An ideal signal is obtained for a feedback level of 0.05 and an external cavity length of 0.30 m. The corresponding spectrum has a compression bandwidth of 22.12 GHz.

In addition, a full analysis of the ambiguity function reveals that the optimized chaotic electric field has a "thumbtack shape" which is ideal for extracting range and Doppler information. A range resolution of 6.7 mm and a pedestal level −30 dB is achieved. For illustrative purposes, the signature of a Boeing 777 is modeled using hot spots, i.e., scattering points.

6.1 ▪ Introduction

The fundamental question we seek to address is whether a semiconductor laser system can produce a chaotic electromagnetic electric field with nearly ideal characteristics for range-Doppler imaging applications. The working hypothesis is that the feedback of the system, which is parameter dependent, can be adjusted to generate this chaotic electromagnetic field, which is the transmitted signal of interest. More specifically we seek to optimize a set of parameters that will produce a chaotic signal that upon processing yields optimum resolution in slant and cross range with minimal sidelobe interference.

6.1.1 ▪ Previous Work and Motivation

Ever since the first implementation of semiconductor lasers in 1962 [1], their optoelectronic characteristics have been widely studied. In 1980, Lang and Kobayashi discovered that by placing a partial mirror in front of a semiconductor laser as a mechanism for external optical feedback, its intensity begins to pulsate chaotically [2]. Lang and Kobayashi developed the theoretical model that is now used to study this type of behavior. For instance, Lenstra et al. showed that the apparent temporal random behavior of the electromagnetic field forces the spectrum to spread its bandwidth well into the GHz range [3]. Masoller performed an analysis of the attractor and bifurcation diagram of the semiconductor laser system with feedback as the feedback level increases [4]. Ye et al. demonstrated that the coherence collapse state can be achieved via the period doubling route or by using a quasiperiodic route to chaos [5]. He and his colleagues performed an experiment to prove that the quasiperiodic route to chaos can be achieved when the oscillation frequency is not equal to a multiple of the external resonant frequency [6]. Kikuchi et al. proved that the route to the coherence collapse state is dependent on the external feedback reflectivity [7]. Mork et al. performed a detailed analysis on a semiconductor laser system [8] and demonstrated that the response depends on a characteristic parameter C that is proportional to the product of the feedback level and on the external cavity length. The parameter C is given by

$$C = \frac{x\tau_{\text{ext}}}{\tau_{\text{in}}}\sqrt{1+\alpha^2}, \tag{6.1}$$

where x is the feedback level, τ_{ext} is the external cavity round trip time, τ_{in} is the internal cavity round-trip time, and α is the line width enhancement factor. In the case of $C < 1$, the frequency response of the laser output is essentially a spectral line. For $C > 1$, the frequency response of the laser is composed of two or more modes. For $C \gg 1$, the frequency response becomes unstable, entering into the coherence collapse state and producing a chaotic signal.

The potential use of chaotic lasers in LADAR applications has yet to be fully explored. Pethel and Corron, using an analog correlator, demonstrated that the temporal correlation of the chaotic laser drops quickly, thus allowing for high resolution range measurements [9]. However, the characterization of the chaotic signal was limited to an experimental time sequence analysis. More recently, a chaotic LADAR was developed by Lin and Liu, using a chaotic laser as the source. The chaotic signal was generated by using an optically injected semiconductor laser. They implemented detection and ranging using temporal correlation [10]. To date there is no evidence in the literature that a full Doppler analysis has been conducted.

6.1.2 ▪ Goals and Methodology

As stated above, our working hypothesis is that the control parameters of the equations that govern the behavior of a semiconductor laser system with external optical feedback can be fine tuned to obtain high resolution in range and Doppler. For this purpose, we propose to perform a full analysis of the behavior of the laser's state variables as a function of two key control parameters, namely the external resonator spacing and mirror reflection coefficient. The complete characterization of the state variables, i.e., electromagnetic field amplitude and phase and population inversion, will serve to prove that the waveform has a chaotic behavior. This characterization includes an in-depth study of time iterations and strange attractors. Furthermore, the ergodicity of the system will be tested for infinitesimal variations in state variable initial conditions. We further propose to adjust and optimize the control parameters based on measurements of the Lyapunov exponent and entropy of the signal. Since the signal processing requires a Michelson interferometer at the receiver, we intend to model the matched filter as an optical correlator to obtain the simulations of autocorrelation and ambiguity function of the field and thus determine the system's performance in the range and Doppler plane. Given the nature of the chaotic signal, this will include a statistical analysis of the ambiguity function. Finally, and for illustrative purposes, imagery of a multiple scattering point target will be generated by using the optimized chaotic signal.

6.2 ▪ Theoretical Background

6.2.1 ▪ Chaotic Signals

The time behavior of a deterministic system is said to become chaotic when it evolves from a periodic to an aperiodic and apparently random behavior. Chaotic systems can be mathematically modeled by differential equations or, in the case of discrete systems, by difference equations. The governing equations of a chaotic system must be nonlinear and include at least three state variables. Otherwise, chaos cannot be induced. A solution of a difference equation is regarded as a sequence of iterates of some initial point [11]. Three descriptors are needed to characterize a chaotic system: the time evolution equations, the values of the parameters describing the systems, and the initial conditions. Chaotic systems are highly sensitive to initial conditions; a very small variation in initial condition produces a rather different signal. In addition, a small change in the control parameters can cause chaos to disappear or to reappear. The chaotic behavior of the system is quantified in terms of its Lyapunov exponents, which measure the degree of divergence of the state space orbit of the system when it forms a strange attractor. Strange attractor is the geometrical manifestation of chaos in a nonlinear dynamical system.

Over the past two decades, chaotic signals have gained significant attention for telecommunication systems applications [12]. Specifically in radar applications, chaotic-based generators are valuable alternatives for testing the vulnerability of radar systems to jamming signals. For instance, one-dimensional chaotic maps have been used to generate quasi-Gaussian signals for microwave radar applications [13, 14]. The effort was directed at finding the value of a control parameter that minimizes the sidelobes of the signal's temporal autocorrelation. In another study, Flores considered several one-dimensional maps to construct a frequency modulated signal with ideal ambiguity function properties [15]. The ambiguity function of the chaotic frequency-modulated (FM) signal was comparable to that of a Gaussian FM signal, which is often used as a standard of performance for random FM radar systems.

Given the ever increasing demands for higher range resolution capability and the need to deploy portable, low-power detection and ranging systems, it is relevant to extend the study of chaotic systems well into the infrared or optical region of the electromagnetic spectrum. Of course, the analysis of chaotic laser systems becomes a priority. To date, there are three different schemes that can be used to generate a chaotic signal: optical injection, optoelectronic feedback, and optical feedback [16, 17, 18]. Owing to the simplicity of the optical-feedback semiconductor system model we choose this approach for proof of concept.

6.2.2 ▪ Semiconductor Laser with Feedback

Semiconductor lasers are commonly used in optical applications since they offer a multitude of significant advantages, including small size, wavelength availability, power, and low cost. These lasers are often classified according to the wavelength of operation and the semiconductor elements that make up the cavity. For instance, a popular semiconductor laser is made of GaAlAs and produces a wavelength from 500 nm to 1000 nm [19].

The semiconductor laser with external optical feedback is implemented by using a laser diode and a partial reflector (mirror) arranged as shown in Figure 6.1 [2].

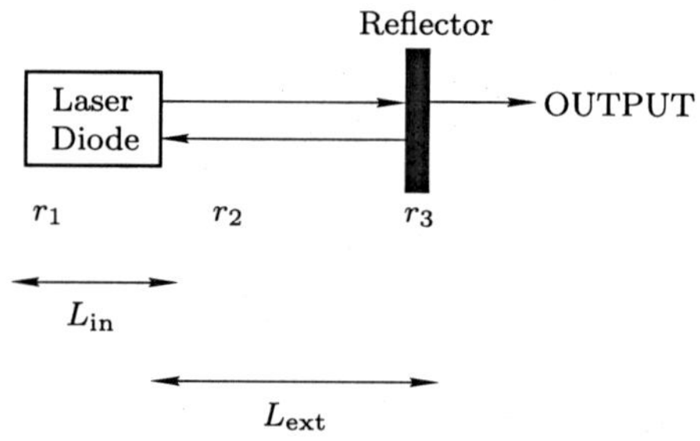

Figure 6.1. *Semiconductor laser with feedback* [29].

The laser diode cavity walls have internal reflectivities r_1 and r_2, while the mirror placed at a distance L_{ext} from the laser has a reflectivity r_3. A fraction of the electric field $E(t)$ that exits the laser is reflected back and enters the laser cavity. The reflected wave has an electric field $E(t-\tau_{ext})$, where $\tau_{ext} = 2L_{ext}/c$ is the external round trip.

Since $E(t-\tau_{ext})$ is out of phase and interferes with the coherence of the field inside the resonator, the total field is strongly dependent to the feedback level, which depends on the reflectivity of the external mirror. The total field falls in several regimes, depending on the reflectivity of the external mirror k measured as a percentage [20]. For very small feedback ($0 < x < 1$), the laser is under single mode operation. For small feedback ($x \cong 1$) the laser undergoes mode hopping. For high feedback ($x > 10$), the laser becomes coherent again and the linewidth is narrowed. Our interest is in the moderate feedback regime ($1 < x < 10$), which produces a chaotic field with fluctuations in the time domain that can increase the laser linewidth from a few MHz up to 50 GHz [3]. This coherence collapse is due to changes in the phase velocity of the waveform inside the cavity. Another way of obtain the range of the different regimes is by using the characteristic parameter C. So, for a fixed value of the external cavity length, the characteristic value C can be obtained for different values of the feedback level.

The electric field and the carrier density of the system are modeled by using the Lang–Kobayashi equations that are obtained from the laser rate equations by adding the feedback term [3]. Thus the resulting state equations are

$$\frac{dE(t)}{dt} = \frac{1}{2}(1+j\alpha)G_N(N(t)-N_{th})E(t) + \frac{\varkappa}{\tau_{\text{in}}}e^{-j\omega_0\tau_{\text{ext}}}E(t-\tau_{\text{ext}}), \tag{6.2}$$

$$\frac{dN}{dt} = J - \frac{N}{\tau_s} - \left[\frac{1}{\tau_p} + G_N(N-N_{th})\right]|E|^2. \tag{6.3}$$

Here $E(t)$ is the complex envelope of the electric field and $N(t)$ is the carrier density. Typical values for physical constants and parameters are listed in Table 6.1 for reference. The noise induced in this system is mainly produced by spontaneous emissions of randomly excited atoms. This effect of such noise is observed in the amplitude and phase of the field [21]. In (6.2), $F_E(t)$ is added as the representation of the Langevin noise associated with the electric field. Likewise, the term $F_N(t)$ is added to (6.3) and will have an impact on the carrier density. For the purposes of this study, all the terms of the Langevin noise are neglected.

Table 6.1. *Parameters of Lang-Kobayashi equations.*

Symbol	Quantity	Value
τ_p	Photon lifetime	4.5 ps
τ_s	Carrier lifetime	700 ps
τ_{ext}	Round trip time in external cavity	Variable
τ_{in}	Round trip time in laser cavity	3.9 ps
$\varkappa^2$	Power reflected from the external cavity	Variable
α	Line width enhancement factor	5
J	Pumping term	2.38706×10^{27}
ω_0	Angular frequency of solitary laser	2.27×10^{15} rad.
N_{th}	Threshold carrier density of solitary laser	1.5×10^{18}
G_N	Differential gain	2.6×10^{-6}

The Lang–Kobayashi dynamical system can be tailored to explicitly show three state variables equations by specifying the amplitude and phase terms of the complex envelope of the electric field:

$$E(t) = A(t)e^{j\phi(t)}. \tag{6.4}$$

Thus,

$$\frac{dE(t)}{dt} = \frac{d[A(t)e^{j\phi(t)}]}{dt} = e^{j\phi}\frac{dA(t)}{dt} + jAe^{j\phi}\frac{d\phi(t)}{dt}. \tag{6.5}$$

Making the corresponding substitution in (6.5) and manipulating its real and imaginary parts will result in the following pair of differential equations:

$$\frac{dA(t)}{dt} = \frac{1}{2}G_N(N(t)-N_{th})A(t) + \frac{\varkappa}{\tau_{\text{in}}}A(t-\tau_{\text{ext}})\cos(\phi(t)-\phi(t-\tau_{\text{ext}})+\omega_0\tau_{\text{ext}}), \tag{6.6}$$

$$\frac{d\phi(t)}{dt} = \frac{1}{2}\alpha G_N(N(t)-N_{th}) + \frac{\varkappa}{\tau_{\text{in}}}\frac{A(t-\tau_{\text{ext}})}{A(t)}\sin(\phi(t)-\phi(t-\tau_{\text{ext}})+\omega_0\tau_{\text{ext}}). \tag{6.7}$$

The system of equations above is a three-dimensional flow where the state variables are the amplitude of the electromagnetic field $A(t)$, its phase ϕ(t), and the carrier density

$N(t)$ [22]. For simulations purposes, the flow is converted to a nonlinear iterated map of the form

$$A(n+1) = f[A(n), j(n), N(n)], \tag{6.8}$$

where $A(n)$ is the discrete amplitude of the electric field phaser, $\varphi(n)$ is its discrete phase, and $N(n)$ is the sampled carrier density which obeys the transformation

$$N(n+1) = g[A(n), j(n), N(n)]. \tag{6.9}$$

Likewise, the discrete version of the phase is of the form

$$j(n+1) = k[A(n), j(n), N(n)]. \tag{6.10}$$

The system can be solved using a fourth-order Runge–Kutta numerical integrator. The sampling frequency has to be chosen to ensure that the chaotic attractor is smooth and that the spectrum of the electromagnetic field is not aliased.

As a note it should be stated that a normalized version of the Lang–Kobayashi equations is commonly found in the literature. The dynamical system is defined in terms of the complex electric field Y and the excess carrier number Z [23] as a function of the normalized time variable s:

$$\frac{dY}{ds} = (1+jR)ZY + \eta Y(s-\theta)\exp(-j\xi), \tag{6.11}$$

$$T\frac{dZ}{ds} = P - Z - (1+2Z)|Y|^2. \tag{6.12}$$

The normalized version of the Lang–Kobayashi equations can be rewritten as a function of three state variables—the electric field Y, the phase φ, and the carrier density Z—by separating the phase of the amplitude in the electric field as follows:

$$\frac{dA_N(s)}{ds} = Z(s)A_N(s) + \eta A_N(s-\theta)\cos(\phi(s) - \phi(s-\theta) + \omega_0\tau_{\text{ext}}), \tag{6.13}$$

$$\frac{d\phi(s)}{ds} = RZ(s)A_N(s) + \frac{\eta A_N(s-\theta)}{A_N(s)}\sin(\phi(s) - \phi(s-\theta) + \omega_0\tau_{\text{ext}}). \tag{6.14}$$

Table 6.2 summarizes the parameters used for the normalized version the Lang–Kobayashi equations which are based on the values listed above.

Table 6.2. *Parameters of normalized Lang-Kobayashi equations* [29].

Symbol	Quantity	Value
η	Feedback rate	Variable
P	Injection current	1
τ_{ext}	Optical feedback delay	variable
τ_p	Photon lifetime	4.5ps
τ_s	Carrier lifetime	700ps
R	Linewidth enhancement factor	5
T	Carrier to photon lifetime	155.5

6.3 ▪ Characterization of Chaotic Signal

6.3.1 ▪ Time Domain and Pseudophase Maps

Chaotic time series have apparently random behavior with no visible periodicity in the time domain. Although much more evidence than the time behavior is needed to confirm that a series is chaotic, time domain plots are a good method for inspecting the behavior of the series.

For simulation purposes, the Lang–Kobayashi equations are solved using a fourth-order Runge–Kutta numerical integrator. The feedback level k is easily adjusted to verify the behavior of the electric field in the periodic, mode hopping, and chaotic regimes of the system. For simplicity, the results are described in terms of the normalized parameters of the Lang–Kobayashi equations. Of course, the same results can be obtained with (6.2)–(6.7).

After eliminating the samples corresponding to the transit time of the external cavity, a time series 0.8 ns in length is obtained. As expected, in the absence of any feedback, simulations show that the complex envelope of the field is constant. Next, the feedback level is set to $\eta = 0.001$. The resulting periodic field of the laser output is observed in Figure 6.2. The electric field has a periodic behavior and is shaped by a single harmonic. The period of the periodic signal depends on the feedback level. The period is inversely proportional to the feedback level, and as the feedback level increases, the period of the signal decreases. However, the period is also directly proportional to the frequency of oscillation, so as the feedback level increases, the oscillation frequency increases. The frequency of oscillation can be determined by using the power spectrum of the signal.

By increasing the feedback level, the signal is forced into the mode hopping regime and is no longer monotone. Instead, several frequency components should appear. As the

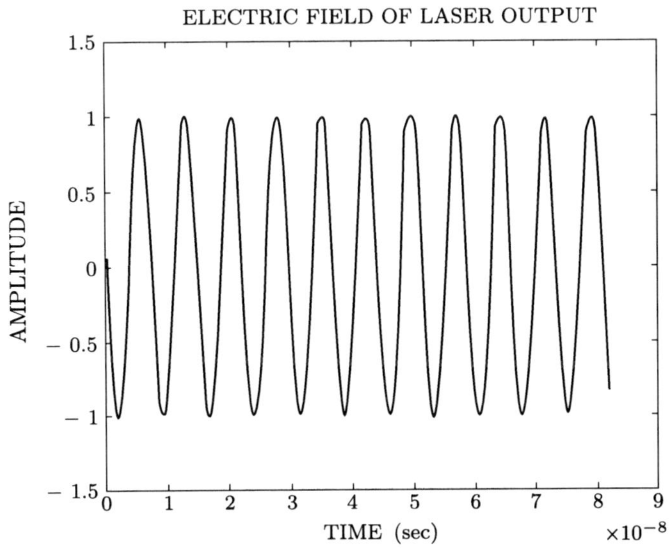

Figure 6.2. *Time behavior of periodic electric field.*

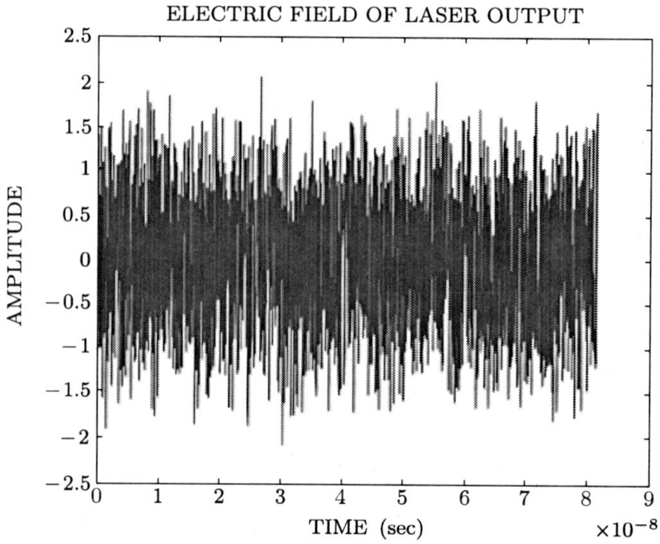

Figure 6.3. *Time behavior of chaotic electric field.*

feedback level is increased to 0.06, the electric field in the time domain becomes chaotic. The time domain behavior is shown in Figure 6.3. Here, the signal has apparently random behavior in time. A detailed analysis of the frequency components of the signal will be provided in the following section.

It is worth mentioning that the examples above illustrate the evolution of the electric field, but neither the chaotic behavior nor the fractal nature has been demonstrated. Although the coherence of the system has collapsed, the fundamental question of whether the field is chaotic or not has yet to be answered. This requires an analysis of the state space orbit.

In a multidimensional system such as the Lang–Kobayashi set of equations, the chaotic performance can be directly inferred by plotting the state variables in a three dimensional space and observing the orbit of the system as it evolves into a strange attractor as will be described in the next section. Alternatively, diffeomorphism can be used when data from only one of the state variables are available [24]. In this instance, a version of the attractor can be constructed by considering delayed versions of the state variable. For instance, since the electric field is the only observable variable, its attractor can be generated by looking at $E(t)$ versus $E'(t)$ or $E(n)$ versus $E(n+1)$.

The plot of $E(n)$ versus $E(n+1)$ is commonly referred to as a pseudophase diagram. The relationship between samples can be distinguished by using this plot. For a better understanding of pseudophase maps and for comparison the pseudophase map of a random signal, a periodic signal and a chaotic signal are shown. For comparison purposes, the pseudophase map of a random signal was generated using a uniform random generator. The pseudophase map of the electric field of the laser was obtained for $\eta = 0.001$ and for $\eta = 0.06$. In the case of the random field, the pseudophase map is composed of many points that are distributed uniformly and do not have any trajectory or relationship, as shown in Figure 6.4. For the periodic field, the pseudophase map has only one orbit reminiscent of the Lissajou pattern commonly used to characterize oscillating electronic circuits. The resulting plot is shown in Figure 6.5. In contrast, the pseudophase map of the chaotic signal has several orbits, and a complex nonrepeated behavior is shown in Figure 6.6.

The pseudophase maps of the carrier density and the phase are shown in Figures 6.7 and 6.8, respectively. In Figure 6.7, it can be observed that there is relationship between the current and next sample of the carrier density. The pseudophase map has a complex behavior that consists of several orbits that do not repeat. This clearly illustrates that there

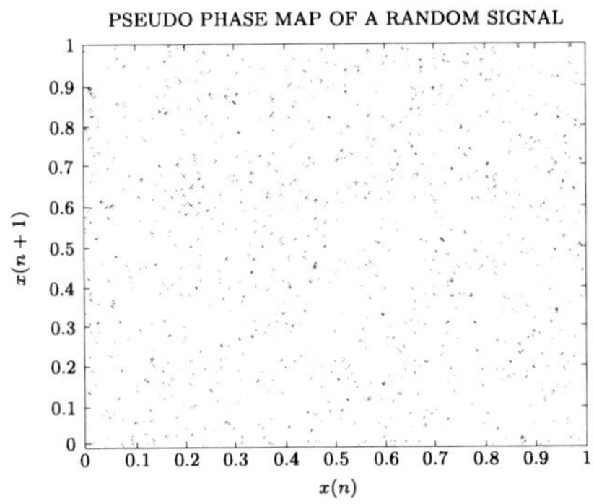

Figure 6.4. *Pseudophase map of a random field.*

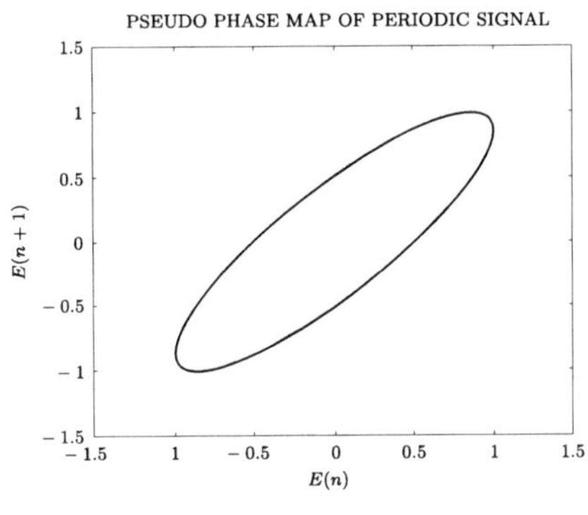

Figure 6.5. *Pseudophase map of periodic field.*

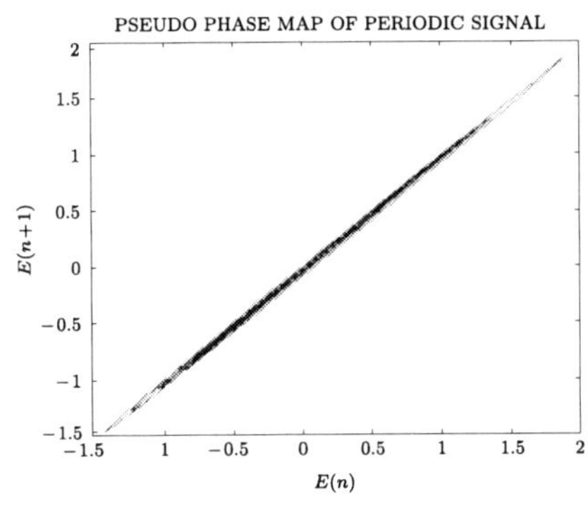

Figure 6.6. *Pseudophase diagram of chaotic field.*

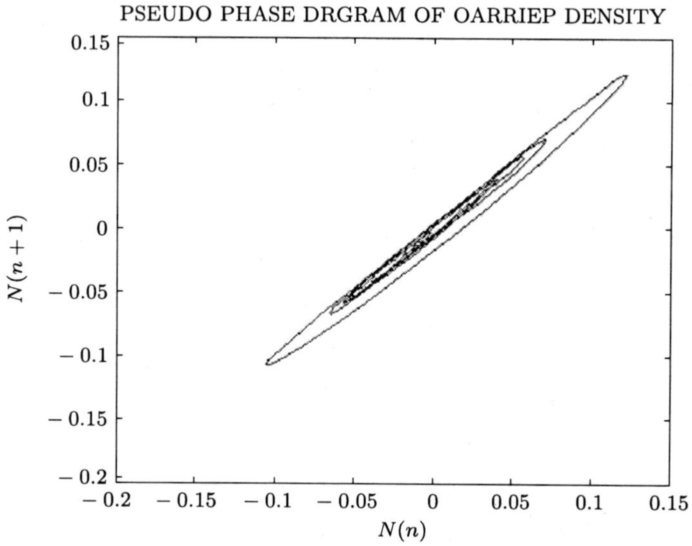

Figure 6.7. *Pseudophase map of carrier density in chaotic regime.*

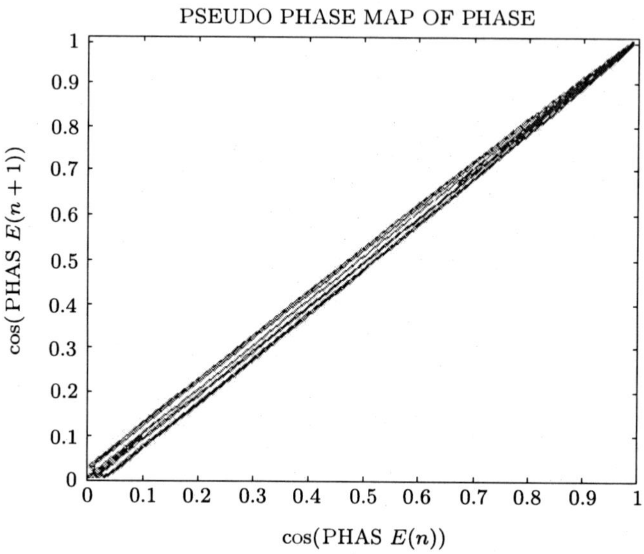

Figure 6.8. *Pseudophase map of carrier density in chaotic regime.*

is a relationship between samples that is confined to a strange attractor. In both cases the pseudophase maps are those of a chaotic signal.

6.3.2 ▪ Attractor of the System and Lyapunov Exponent

The attractor defines the behavior of the system variables. Specifically, for a semiconductor laser with feedback the attractor is a three-dimensional plot of the carrier density, the electric field of the laser output, and its phase. In Figure 6.9 the attractor of the system for $\eta = 0.001$ can be observed. In this case, the attractor of the system consists of just one orbit (or limit cycle) because the orbit follows a periodic pattern.

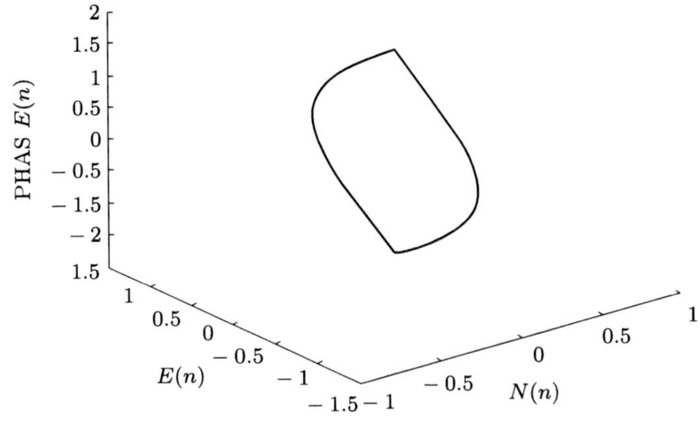

Figure 6.9. *Attractor of the system in the periodic regime* [29].

As the feedback level is increased to 0.01 the signal enters into a quasi-periodic regime or mode hopping regime. In this case the attractor of the system follows what is called a torus form. The attractor of the system in the mode hopping region is shown in Figure 6.10.

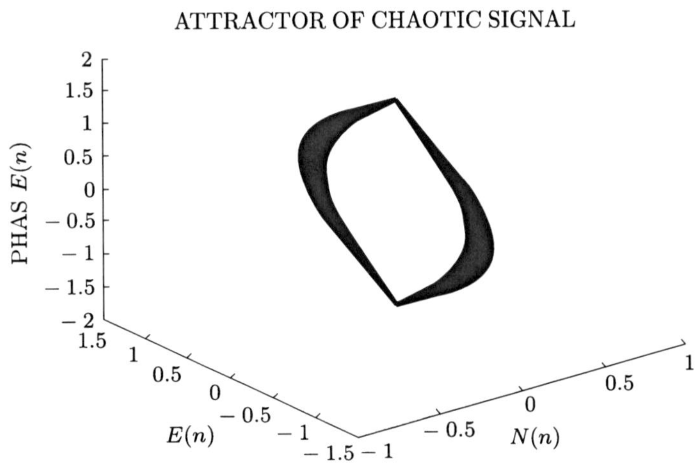

Figure 6.10. *Attractor of the system in the mode hopping region* [29].

For $\eta = 0.06$ the system becomes chaotic, as shown in Figure 6.11. In this case, the attractor follows a complex but deterministic trajectory.

The Lyapunov exponent measures the divergence of a system's state orbit due to small changes in the initial conditions. A system with three variables will have three Lyapunov exponents $\lambda_1 > \lambda_2 > \lambda_3$ [24]. The largest Lyapunov exponent λ_1 determines whether a system is chaotic. For the field of the semiconductor laser to be chaotic, the largest Lyapunov exponent must be positive. In the event that the largest Lyapunov exponent

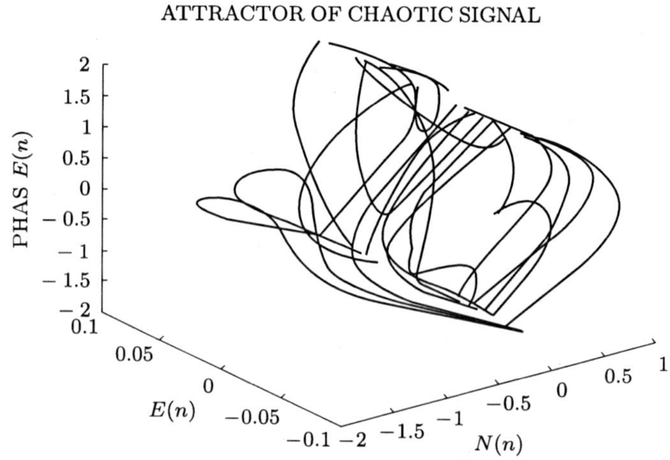

Figure 6.11. *Attractor of system in the chaotic regime* [29].

is less than zero, the electric field approaches an equilibrium point. When the largest Lyapunov exponent is near zero, the system will approach a periodic orbit or limit cycle.

For high-order maps the Lyapunov exponent can be approximated as [24, 25]

$$\lambda = \frac{1}{n}\ln\left(\frac{d_n}{d_o}\right), \tag{6.15}$$

where d_o is the initial distance between the initial points of two neighboring orbits O_i and O_j,

$$d_o = |\rho_j - \rho_i|, \tag{6.16}$$

and d_n is the distance between the orbits for the nth iteration,

$$d_n = |\rho_{j+n} - \rho_{i+n}|. \tag{6.17}$$

In the particular case of three-dimensional maps, the Lyapunov exponent can be calculated by using the procedure developed by Sprott [24].

In the case that the dynamical equations of the system are not known, the Lyapunov exponent can be calculated by using the "poor man's Lyapunov exponent," which is equal to the inverse of the width of the main lobe of the autocorrelation function [24].

The largest Lyapunov exponent was calculated for the four regimes of the laser output. Table 6.3 shows the Lyapunov exponent calculation for each regime. In the chaotic regime the Lyapunov exponent is greater than in the other three regimes. As anticipated, in the periodic regime, the Lyapunov exponent is close to zero.

6.3.3 ▪ Time Series Properties

In previous sections the Lang–Kobayashi equations were solved by setting the initial conditions to fixed values. The control parameters in this case can be modified so that the desired signal is obtained. However, when a LADAR receiver captures a signal that is returning from a target, the original conditions are not necessarily known by the receiver. In this case, the receiver captures only the experimental data or the time series

Table 6.3. *Lyapunov exponents for different regimes.*

Regime	x	η	λ_1
Periodic	0.0008	0.001	0.0395
Mode Hopping	0.008	0.01	0.1131
Chaotic	0.052	0.06	0.3208
Periodic	0.433	0.5	0.0127

data. Therefore, it is important to analyze the time series properties and discover if the statistical properties change with time. In the case that the statistical properties are the same, whether they are calculated from a probabilistic approach or via a temporal time analysis, the system is said to be ergodic.

In order to perform the temporal time analysis, a total of 100,000 samples of the electric field are used. In this case, the laser output is set in the chaotic regime by setting the feedback level at 0.06 and the external cavity length to 0.09 m. To perform the equivalent probabilistic analysis the experiment is run 1000 times. Each time the initial conditions of the Lang–Kobayashi equations are changed. The electric field of the 10,000th sample in each run is saved.

Next, the first four moments are calculated for the temporal and ensemble electric field vectors. The four moments calculated are the mean, standard deviation, skewness, and kurtosis.

The kurtosis for a Gaussian distribution is zero. Therefore, if a kurtosis is positive, the distribution is called leptokurtic or flat tailed; if the kurtosis is negative, the distribution is called platykurtic.

To test the ergodicity of the field, a vector consisting of 100,000 iterations is created. The four moments are calculated for this time series. Results are shown in Table 6.4. The mean of the electric field is almost zero; the standard deviation is close to 1; the mass is concentrated on the left side, based on the skewness; and the distribution is platykurtic based on the kurtosis.

Table 6.4. *Probabilistic and temporal statistical properties of electric field.*

Parameter	Probabilistic	Temporal
Mean	0.005	0.002
Standard Deviation	0.7112	0.719
Skewness	−0.027	−0.016
Kurtosis	−0.6123	−0.588

The probability distribution of the amplitude of the electric field is shown in Figure 6.12. The probability distribution has a quasi-Gaussian shape.

The probability density function of the 10,000th sample of the electric field is shown in Figure 6.13.

The corresponding estimates of the four moments are very close to each other. Furthermore, the probability distributions are similar. Hence, the electric field can be assumed to be ergodic.

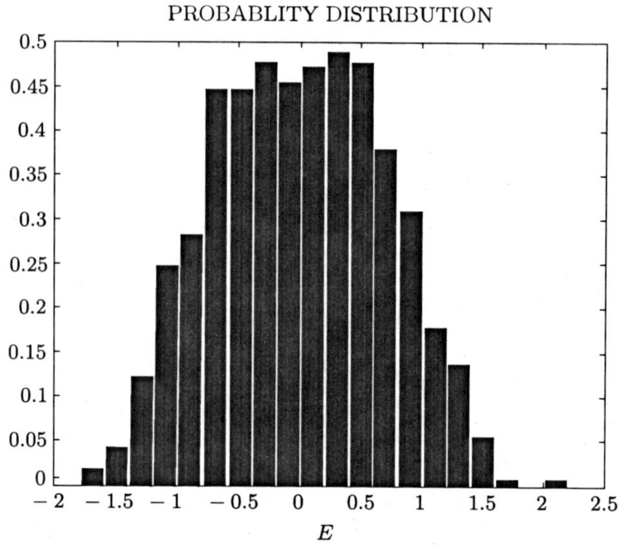

Figure 6.12. *Probability density of electric field based on one realization.*

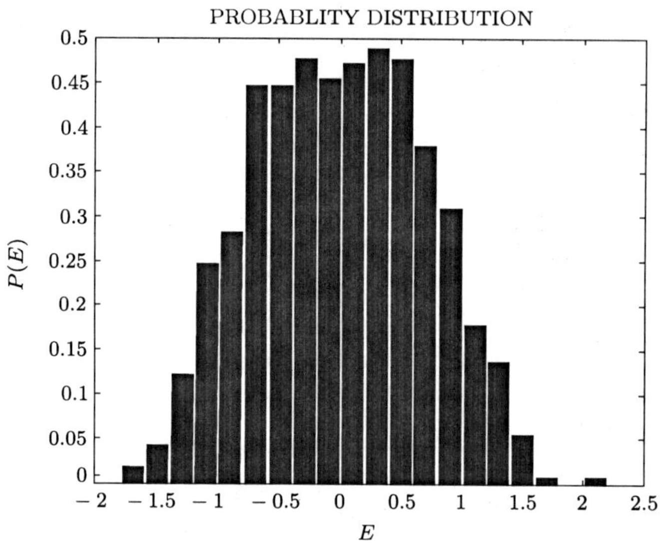

Figure 6.13. *Probability density of nth sample of the electric field.*

6.3.4 ▪ Power Spectrum and Autocorrelation

To analyze the impact that the control parameters have on the power spectrum of the electric field, the bifurcation diagram of the power spectrum is obtained. Figure 6.14 shows the bifurcation diagram of the power spectrum for a fixed external cavity length and varying the feedback level. For each feedback level the power spectrum of the electric signal is obtained and plotted versus the frequency. For illustration purposes, the power spectrum was normalized to its maximum value. The power spectrum of the electric field has a dependence on the feedback level. As the feedback level increases, more frequencies appear on the power spectrum. At low feedback level, the power spectrum is composed of just one frequency. As the feedback level is increased, more frequency components appear on the power spectrum until chaos where many frequency components are present is reached.

In addition, the influence that the external cavity length has on the power spectrum was analyzed by maintaining the feedback level fixed and varying the external cavity length. Figure 6.15 shows the bifurcation diagram of the power spectrum varying the external cavity length from 0.00001 m< L_{ext} < 0.30 m. It can be observed that the power

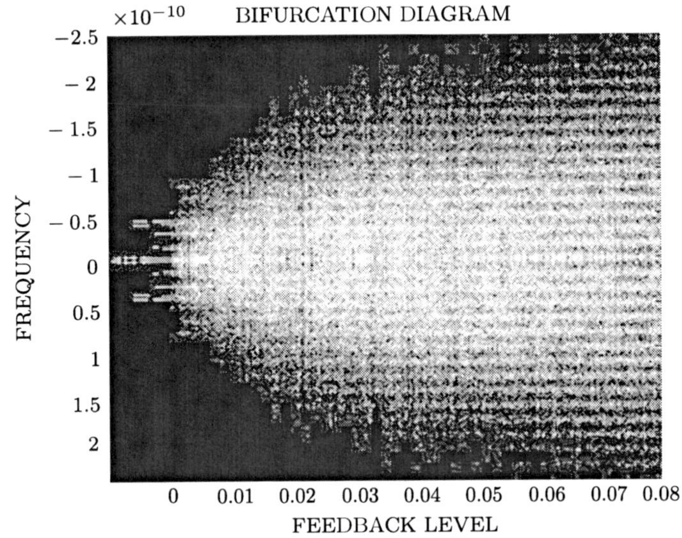

Figure 6.14. *Bifurcation diagram for a fixed external cavity length.*

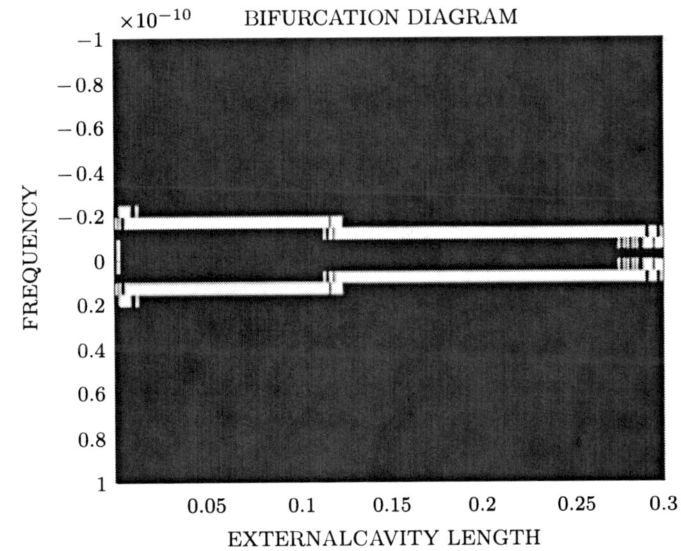

Figure 6.15. *Bifurcation diagram for a fixed feedback level.*

spectrum of the electric field does not go to another regime as the external cavity length is changed. The power spectrum of the signal is periodic.

A primary objective is to generate a chaotic electromagnetic field that has a flat, wide spectral distribution. The power spectrum is calculated by obtaining the magnitude square of the Fourier transform of the time domain field [24].

After, eliminating the transient samples, a total of 4096 samples are used. An average of 10 power spectra is obtained to have a more accurate result. The power spectrum of the electric field for the periodic regime is shown in Figure 6.16. In this case, the power spectrum has one frequency component. This means that the signal is periodic. The same power spectrum is obtained by setting the feedback level in the range of $0 < x < 0.01$. In that feedback level regime, just one frequency is present. However, the frequency is dependent on the feedback level. As the feedback level increases, the frequency of the periodic signal increases.

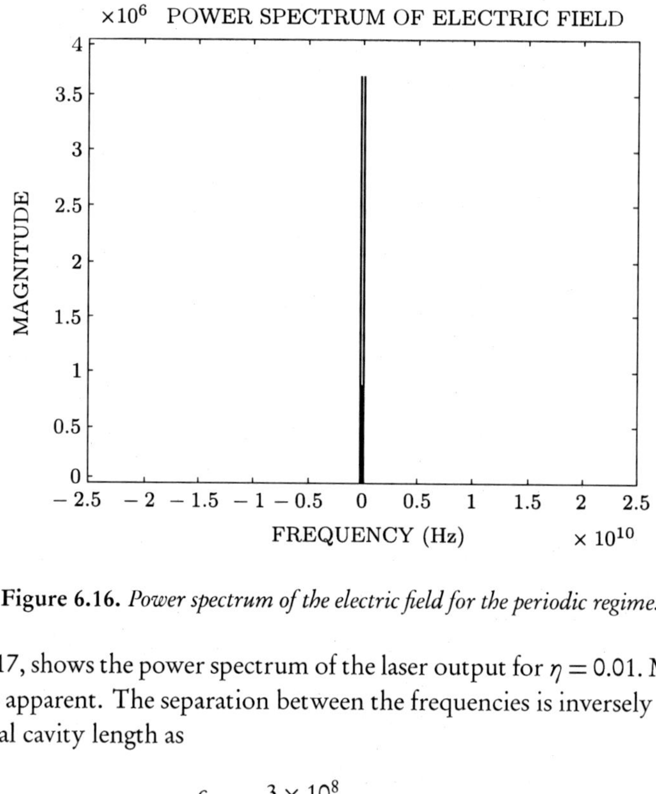

Figure 6.16. *Power spectrum of the electric field for the periodic regime.*

Figure 6.17, shows the power spectrum of the laser output for $\eta = 0.01$. Many spectral lines become apparent. The separation between the frequencies is inversely proportional to the external cavity length as

$$f_{\text{ext}} = \frac{c}{2L_{\text{ext}}} = \frac{3 \times 10^8}{2(0.09)} = 0.166 \times 10^{10} Hz. \tag{6.18}$$

The frequencies of the spectral lines can be easily distinguished as multiples of the external frequency f_{ext}. This type of behavior is encountered for a feedback level around $0.01 < x < 0.019$.

Figure 6.18 shows the power spectral density of the laser output for $\eta = 0.06$. There are frequency components in this case, and the bandwidth increases to 22 GHz. The external cavity frequency cannot be distinguished anymore. In this case, the coherence of the laser collapses and the amplitude of the power spectrum is significantly lower than the amplitude of the power spectrum in the periodic regime. In fact, the amplitude of

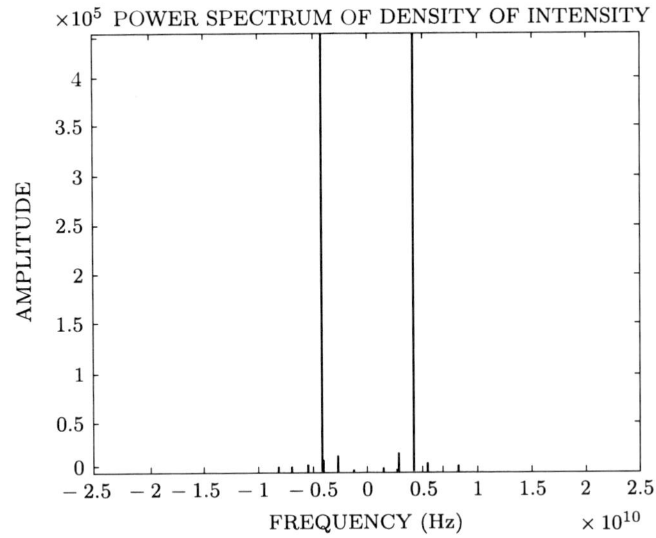

Figure 6.17. *Power spectrum of the electric field for the mode hopping regime.*

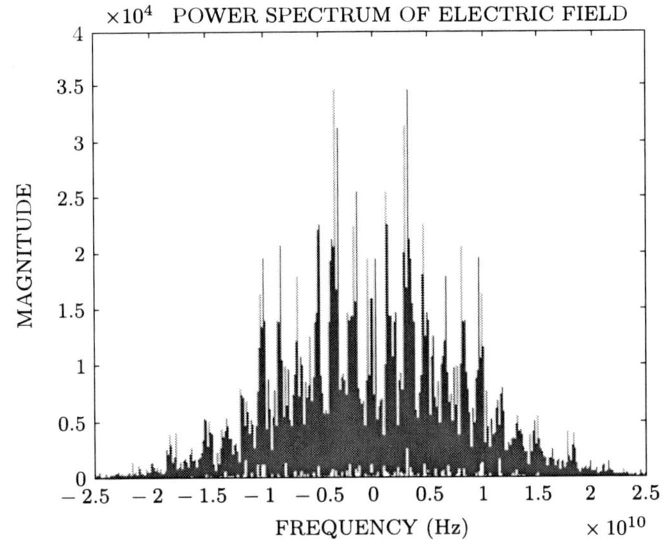

Figure 6.18. *Power spectrum of the electric field for chaotic regime* [29].

the power spectrum in the chaotic regime is 100 times lower than the one of the periodic regime. These type of behavior of the power spectrum is encountered for a feedback level $0.01 < x < 0.10$.

The autocorrelation is obtained by computing the inverse Fourier transform of the power spectrum as shown in equation (6.36) [24].

The autocorrelation of the electric field for $\eta = 0.001$ is shown in Figure 6.19. In this case, the autocorrelation has many sidelobes. As the time lag is varied, the sidelobes of the autocorrelation do not decrease significantly. That is an indication that that the signal is not chaotic or random it is periodic.

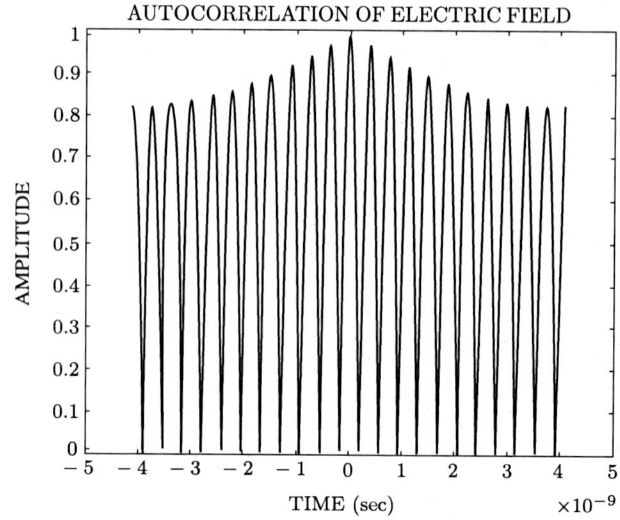

Figure 6.19. *Autocorrelation of the electric field for the periodic regime.*

The autocorrelation of the electric field for a feedback level $\eta = 0.01$ is shown in Figure 6.20. In this case, the same pattern is repeated. This is an indication that the electric field signal has some periodic behavior. The autocorrelation of the signal has a triangular envelope, but the sidelobe levels are quite high.

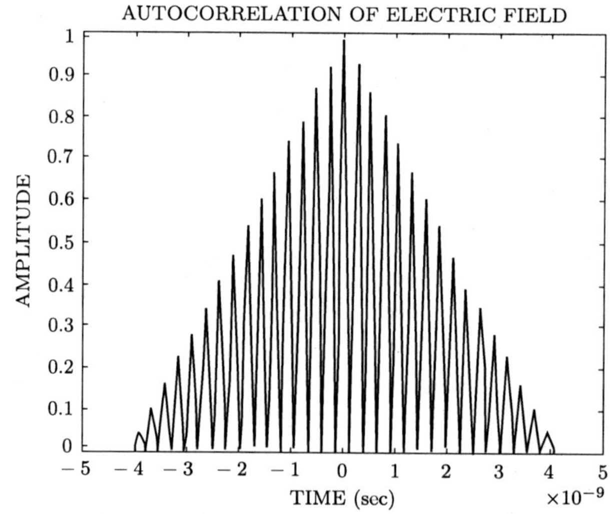

Figure 6.20. *Autocorrelation of the electric field of the mode hopping regime.*

The autocorrelation of the electric field for a feedback level $\eta = 0.06$ is shown in Figure 6.21. In this case the autocorrelation has one peak, and the signal decorrelates with time. The autocorrelation does not repeat itself, which is an indication that no periodic component is present. The sidelobes are less than 0.3. A detail analysis of the field in the chaotic regime will be performed in next section.

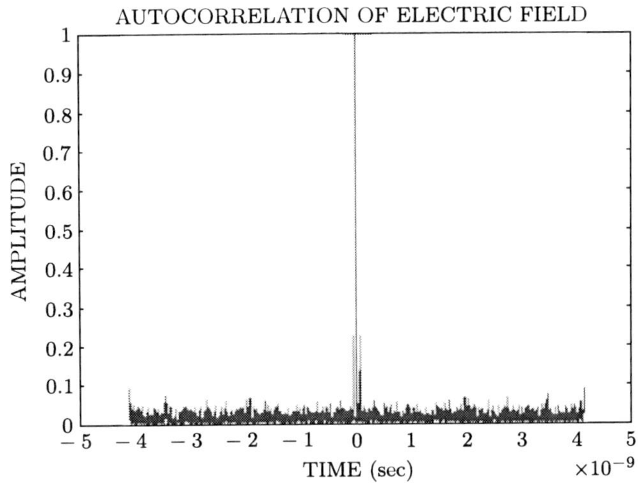

Figure 6.21. *Autocorrelation of the electric field for the chaotic regime.*

6.4 ▪ Optimization of Control Parameters

6.4.1 ▪ Parameter Selection Approach

In ranging applications, the selection of the transmitted signal is a crucial matter since it becomes the reference against which the echo from the target is compared. More specifically, in the detection process, the echo is correlated with the transmitted signal to obtain the distribution of the scattering point of the target in the range-Doppler domain. In this plane, the bandwidth of the transmitted signal determines the range resolution, while its duration determines the Doppler resolution.

There is, of course, a myriad of signals that could be used depending on the target or scenario to be mapped. A signal that has optimal properties can be deterministic, chaotic, or random. Properties for deterministic and random signals have been studied extensively. Those of a chaotic signal are less well known. For this reason, we propose that an ideal chaotic signal must have the following properties:

- ergodicity,
- flat, wide spectrum,
- narrow autocorrelation,
- negligible autocorrelation and ambiguity surface sidelobes,
- low ambiguity surface entropy,
- high Lyapunov exponent.

In what follows we consider the chaotic signal obtained with the optical feedback semiconductor laser described above. It is important to mention that the optimization of the signal is performed in the chaotic regime of the laser output. The other two regimes are not considered.

6.4.2 ▪ Statistical Properties of Chaotic Signal

Although the chaotic signal is strictly deterministic, it can be treated as the realization of a random process, provided that the initial conditions of the laser dynamical system are unknown. In this case, the statistical properties of the chaotic signal become of interest. To start, the first four moments of the chaotic signal are obtained.

Consider 100,000 samples of the chaotic signal generated by setting the feedback level to 0.05. For this experiment, the feedback level is fixed to 0.05, and the external cavity length is varied initially in steps of 0.01 m and then in steps of 0.05 m. Similar tests were performed by using the length of the external cavity as a multiple of the laser's original half wavelength. Table 6.5 shows the corresponding statistical properties of the probability density function of the chaotic signal. Changing the external cavity length does not appear to have a significant impact on the standard deviation of the chaotic signal. However, the mean, the kurtosis, and the skewness do vary somewhat but not in a discernible pattern.

Table 6.5. *Four moments of the chaotic signal for variable external cavity length.*

0.05	−0.019	0.717	0.032	−0.624
0.10	−0.019	0.721	0.036	−0.549
0.15	−0.003	0.716	0.028	−0.542
0.20	−0.003	0.719	0.030	−0.485
0.25	−0.023	0.721	0.004	−0.579
0.30	−0.002	0.718	0.003	−0.545

The next step is to analyze the influence that the feedback level has on the statistical properties of the chaotic signal. For this experiment, 100,000 samples of the chaotic signal were generated by setting the external cavity length to $L_{ext} = 0.3$ m while varying the feedback level between 2 percent and 9 percent in increments of 1 percent. As shown in Table 6.6, the feedback level provides a signal with a mean close to zero is 0.05. In this case, the probability distribution is almost symmetric about the origin having a skewness of 0.003. However, the distribution is not Gaussian since the kurtosis deviates significantly from 3.0.

Table 6.6. *Four sample moments of chaotic signal for various feedback levels.*

0.02	0.056	0.697	−0.067	−1.055
0.03	−0.003	0.715	−0.025	−0.894
0.04	−0.007	0.720	−0.035	−0.594
0.05	−0.002	0.718	0.003	−0.545
0.06	−0.013	0.720	0.027	−0.567
0.07	−0.044	0.725	0.040	−0.528
0.08	−0.020	0.737	0.015	−0.437
0.09	−0.015	0.737	−0.023	−0.418

6.4.3 ▪ Autocorrelation and Power Spectrum for Variable External Cavity Length

For the analysis of the autocorrelation and power spectrum of a signal, a total of N iterations of the Lang–Kobayashi equations are computed. In practice, the first m samples should be discarded to allow the state variables to settle into the strange attractor. Therefore, a total of $N-m$ samples are used to study the chaotic signal. To reduce the fluctuations in the autocorrelation and the power spectrum, M runs of the experiment should be conducted to obtain ensemble averages. In what follows, $N = 6096, m = 2000$, and $N-m = 4096$. This last value is chosen so that the fast Fourier transform computations are done efficiently.

In this case the Lang–Kobayashi equations are solved by setting the feedback level η to 0.05 (5 percent). This value of feedback level is in the chaotic regime. Next, the external cavity length is set to 0.05 m. The autocorrelation of the electric field is shown in Figure 6.22. In this case, the major side lobe level is 0.2132. The major sidelobe level is located at a time equal to the external round trip time given by

$$\tau_{\text{ext}} = \frac{2L_{\text{ext}}}{c} = \frac{2(0.05)}{3\times 10^8} = 0.33\times 10^{-9}\text{sec.} \tag{6.19}$$

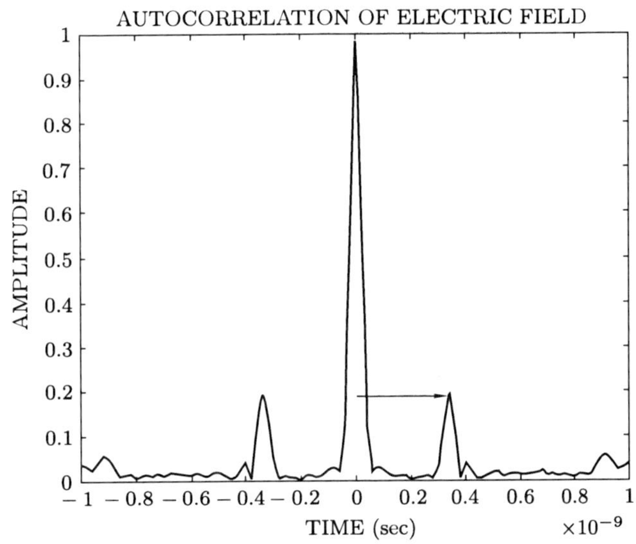

Figure 6.22. *Autocorrelation of the electric field for $L_{\text{ext}} = 0.05$ m.*

The autocorrelation of the electric field for an external cavity length of 0.30 m is shown in Figure 6.23. In this case, the major sidelobe level has a maximum value of 0.1996, and the major sidelobe is located at $\tau_{\text{ext}} = 2\times 10^{-9}$sec.

The inverse of the mainlobe width, also known as the poor man's Lyapunov exponent, and the compression bandwidth of the signal are estimated from the autocorrelation of the chaotic field and are shown in Table 6.7. Here the signal's bandwidth is calculated by taking the reciprocal of the main lobe width of the autocorrelation. Notice that the external cavity length does not have an impact on any of the parameters of the autocorrelation.

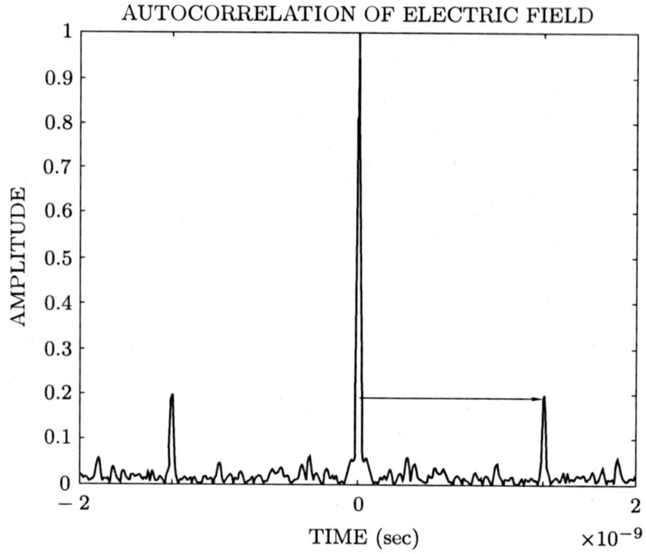

Figure 6.23. *Autocorrelation of the electric field for* $L_{ext} = 0.030$ m.

Table 6.7. *Main lobe width, poor man's Lyapunov exponent (LE), and bandwidth of chaotic field.*

L_{ext}	Main Lobe Width	Poor Man's LE	Compression Bandwidth
0.05	4.64×10^{-11}sec.	0.4310	21.55 GHz
0.10	4.53×10^{-11}sec.	0.4415	22.07 GHz
0.15	4.52×10^{-11}sec.	0.4424	22.12 GHz
0.20	4.47×10^{-11}sec.	0.4474	22.37 GHz
0.25	4.57×10^{-11}sec.	0.4376	21.88 GHz
0.30	4.52×10^{-11}sec.	0.4492	22.12 GHz

The power spectrums of the electric field for an external cavity lengths of $L_{ext} = 0.05$ m and 0.30 m are shown in Figures 6.24 and 6.25, respectively. In the two cases the bandwidth of the power spectrum is comparable.

6.4.4 ▪ Autocorrelation and Power Spectrum for Variable Feedback Level

In the previous section, the effect of varying the external cavity length on the autocorrelation and the power spectrum was analyzed. However, we are also interested in analyzing the effect of the feedback level on the autocorrelation and the power spectrum of the signal.

Thus, the autocorrelation of the chaotic signal is obtained for different feedback levels in the chaotic regime and for an external cavity length of $L_{ext} = 0.30$ m. In Figure 6.26 the normalized autocorrelation of the electric field is shown for a feedback level $\eta = 0.02$, which corresponds to the lower threshold of the chaotic regime. In this case, the highest side lobe level is 46.6 percent of the peak value.

Notice that the sidelobe levels are extremely high, which indicates that the signal does not decorrelate with time. In addition, some degree of periodicity can be observed. This

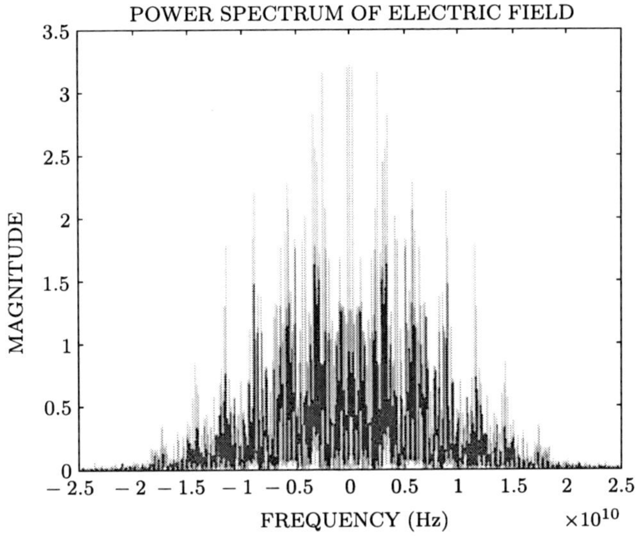

Figure 6.24. *Power spectrum of electric field for* $L_{ext} = 0.05$ m.

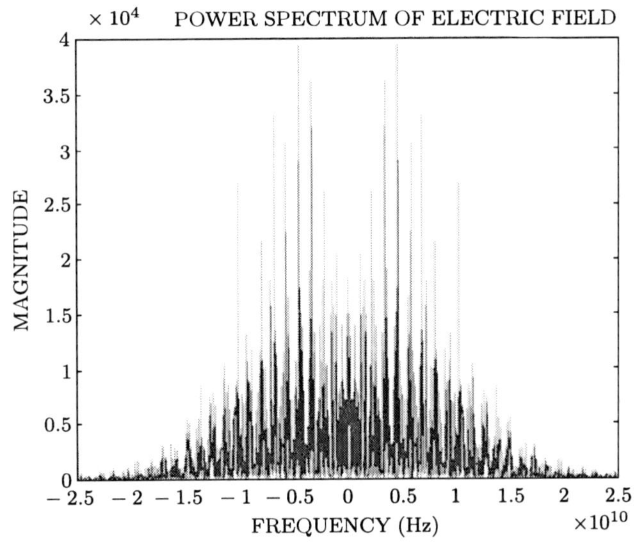

Figure 6.25. *Power spectrum of the electric field for* $L_{ext} = 0.30$ m.

is consistent with the behavior of the strange attractor that forms a toroid around a steady state orbit.

Figure 6.27 shows the autocorrelation of the signal for a higher feedback level of 0.05. In this case, the highest sidelobe level is 0.1828 and the samples are seen to decorrelate quickly.

It is clear that the autocorrelation has just two sidelobes with high amplitude. The other sidelobes are lower than 5 percent of the peak intensity. The location of these sidelobes depends on the length of the cavity, as will be explained below.

Figure 6.28, shows the autocorrelation of the electric field for a feedback level of 0.09, which is the upper limit of the chaotic regime. In this case, the highest sidelobe reaches

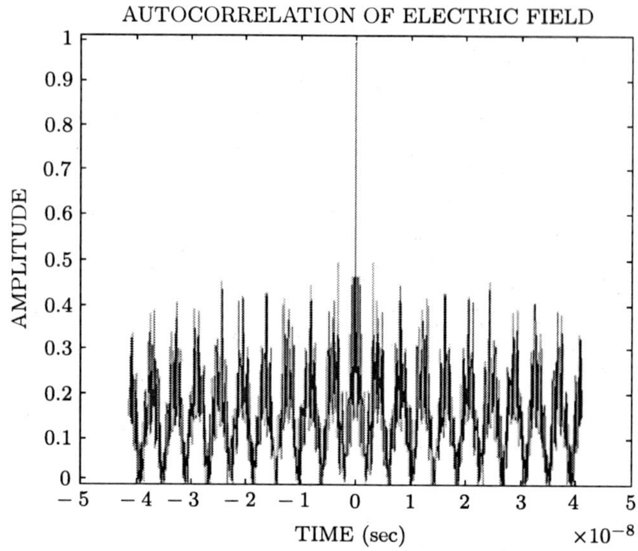

Figure 6.26. *Autocorrelation of the electric field for* $\eta = 0.02$.

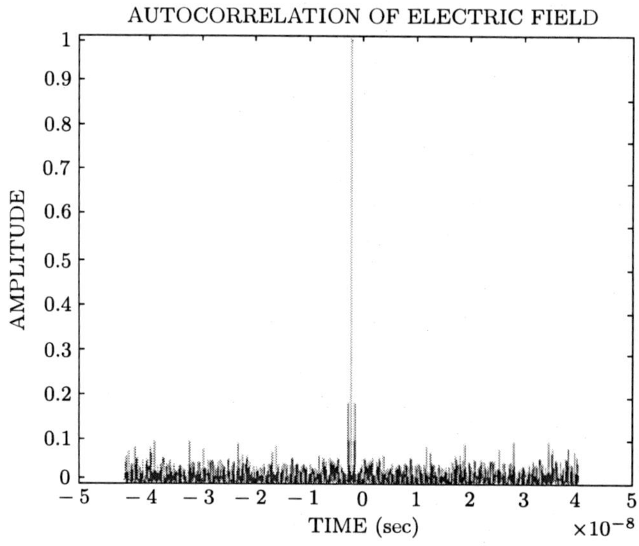

Figure 6.27. *Autocorrelation of the electric field for* $\eta = 0.05$ [29].

a level of 34 percent. Although the samples decorrelate in time, the autocorrelation has four sidelobes.

Figure 6.29 shows a closeup of the autocorrelation of the electric field. In this figure, four sidelobes can be easily distinguished, which would indicate that the signal has regained some periodicity.

The power spectrum of the electric field for a feedback level $\eta = 0.02$ is shown in Figure 6.30. In this case, we observe that the spectrum of the signal is no longer coherent and spreads over a bandwidth of 12.82 GHz.

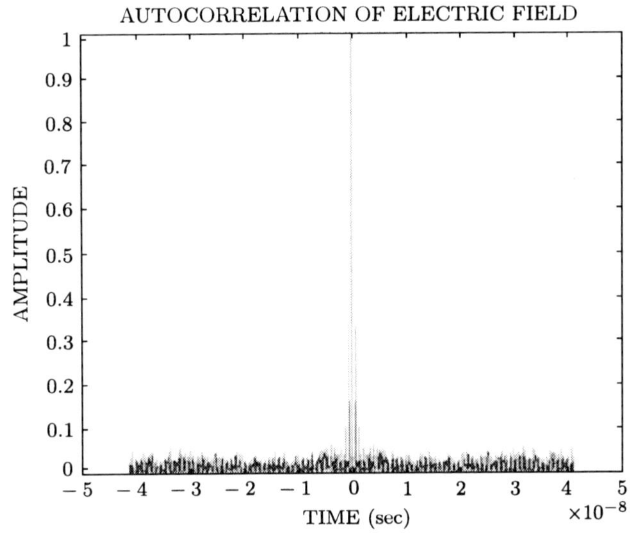

Figure 6.28. *Autocorrelation of the electric field for* $\eta = 0.09$.

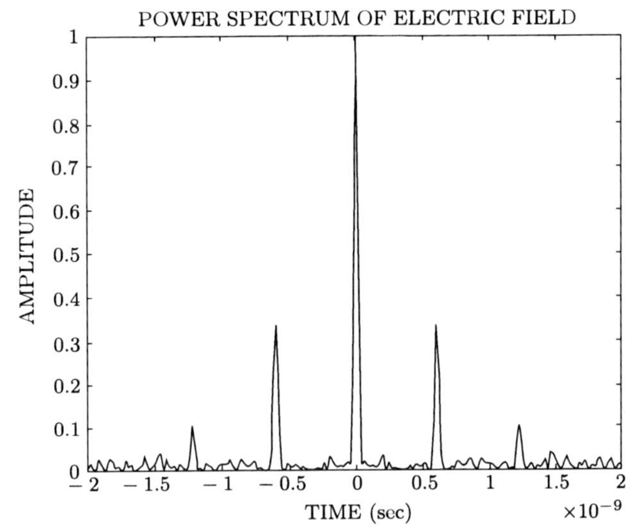

Figure 6.29. *Zoom in of the autocorrelation of the electric field for* $\eta = 0.09$.

The power spectrum of the electric field for a feedback level $\eta = 0.05$ is shown in Figure 6.31. Although the compression bandwidth of 22.12 GHz is close in value to the bandwidth obtained in the previous case, the energy is spread more evenly over a wider frequency range.

The power spectrum of the electric field for a feedback level $\eta = 0.09$ is shown in Figure 6.32. The bandwidth in this case increased to 30.48 GHz by increasing the number of modes in the signal. At this point the laser output has a number of mode lines.

It is clear from these simulations that the spectrum reaches an ideal distribution somewhere near the middle of the range of η. However, the analysis of the spectral distribution alone is insufficient to select the feedback level that gives the best chaotic signal.

Table 6.8. *Autocorrelation main lobe width, poor man's Lyapunov exponent (LE), and bandwidth of electric field for variable η.*

Feedback Level η	Main Lobe Width	Poor Man's LE	Compression Bandwidth
0.02	7.8×10^{-11} sec.	0.2564	12.82GHz
0.03	6.16×10^{-11}sec.	0.3246	16.23GHz
0.04	5.18×10^{-11}sec.	0.3861	19.30GHz
0.05	4.52×10^{-11}sec.	0.4492	22.12GHz
0.06	4.54×10^{-11}sec.	0.4405	22.02GHz
0.07	3.88×10^{-11}sec.	0.5154	25.77GHz
0.08	3.5×10^{-11}sec.	0.5714	28.57GHz
0.09	3.28×10^{-11}sec.	0.6097	30.48GHz

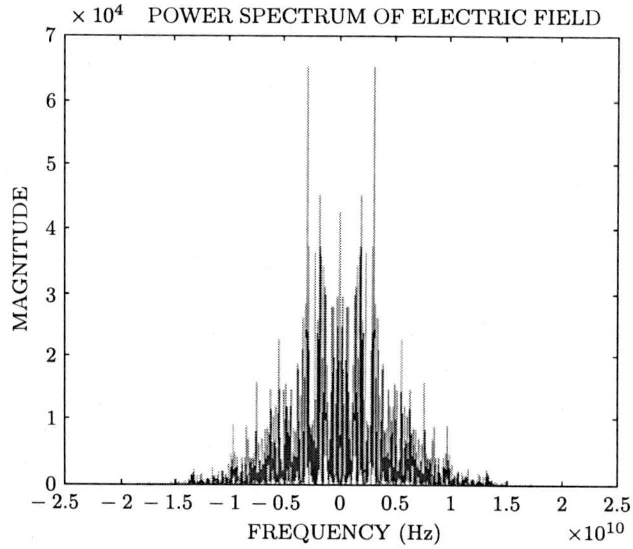

Figure 6.30. *Power spectrum of the electric field for* $\eta = 0.02$.

6.4.5 ▪ Feedback Level Selection

In this section, we seek additional ways to characterize the chaotic electric field generated by the LADAR system under study. For this purpose we consider the entropy of the field's autocorrelation, the Lyapunov exponent of the field, and the effective bandwidth of its spectrum.

The concept of entropy is introduced to evaluate the statistical characteristics of the electric field. In analogy to Shannon's entropy, the entropy of the field's temporal autocorrelation is defined as

$$S_R = -\sum_{i=1}^{N} \frac{|R_E|}{\sum_{i=1}^{N} R_E} \log \frac{|R_E|}{\sum_{i=1}^{N} |R_E|}, \tag{6.20}$$

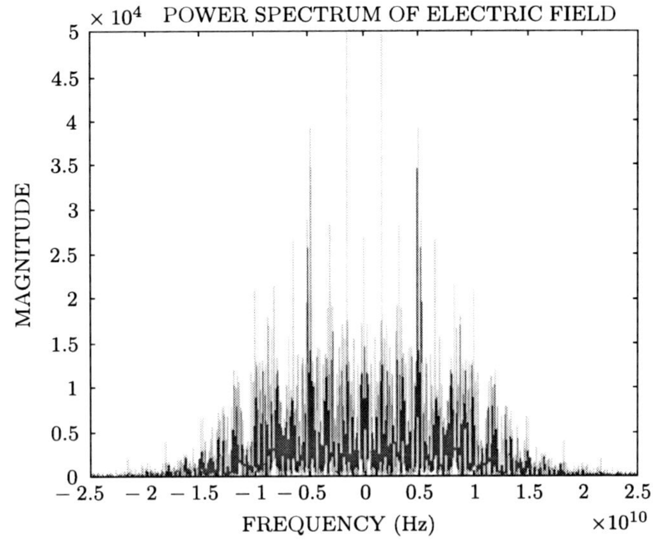

Figure 6.31. *Power spectrum of the electric field for* $\eta = 0.05$.

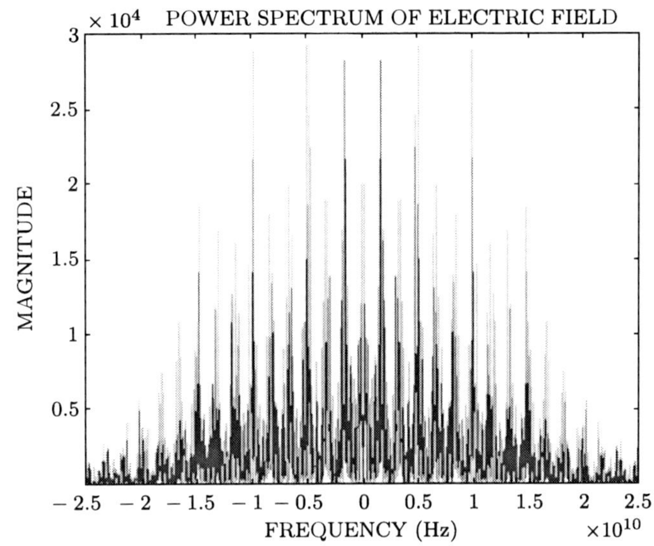

Figure 6.32. *Power spectrum of the electric field for* $\eta = 0.09$.

where R_E is the autocorrelation of the electric field and N is the total number of samples in the autocorrelation. A low entropy of autocorrelation means that the autocorrelation has low disorder; high entropy means more disorder. For a chaotic signal, high entropy indicative of more disorder is anticipated. Similarly, the entropy of the ambiguity surface associated with the field is defined as [27]

$$S_\chi = -\Sigma\Sigma I_{i,j} \log I_{i,j}, \tag{6.21}$$

where

$$I_{i,j} = \frac{\chi_{i,j}}{\Sigma\Sigma\chi_{i,j}} \tag{6.22}$$

and $\chi_{i,j}$ is the ambiguity function where i is the row index and j is the column index. The entropy is a measure that is used to evaluate the degree of randomness of a variable. For instance, an experiment with a strong bias toward a single value has lower entropy compared with another experiment with a wider distribution of values. Thus, we speculate that an entropy value can be associated with the temporal autocorrelation of a signal. The faster the autocorrelation dies out with increasing delay, the lower its entropy. Conversely, if the autocorrelation dies out slowly or shows multiple sidelobes, it must have higher entropy.

In contrast, the Lyapunov exponent is used as a measure of divergence of the electric field from its initial conditions. For instance, consider two sequences of the electric field with slightly different initial conditions. A high Lyapunov exponent value indicates that the electric field orbits quickly diverge from each other. The third parameter used is the effective (root mean square) bandwidth of the field. This is obtained by considering the spectral weighing in the strength of the spectral distribution. A large effective bandwidth means that the power of the signal is widely distributed in the frequency domain. Conversely, a lower effective bandwidth indicates that the power is narrowly distributed.

Table 6.9 summarizes the values of the entropy, Lyapunov exponent, and effective bandwidth obtained through simulations of the electric field for multiple feedback levels. For completeness the entropy of the signal sample distribution is included.

Table 6.9. *Evaluation of electric field for different feedback levels* [29].

η	S(corr)	S(pdf)	LE	SLL	Effective Bandwidth
0.01	3.427	5.130	0.2212	0.466	3.23GHz
0.02	1.194	5.517	0.2956	0.145	4.98GHz
0.03	1.114	5.965	0.2843	0.171	6.67GHz
0.04	0.839	6.058	0.3142	0.175	7.74GHz
0.05	0.742	6.210	0.3281	0.183	8.38GHz
0.06	0.767	6.229	0.3107	0.227	8.35GHz
0.07	0.851	6.298	0.3046	0.282	9.30GHz
0.08	0.929	6.318	0.3223	0.315	10.0GHz
0.09	0.862	6.409	0.3624	0.340	10.4GHz

In previous sections we found that the autocorrelation of the electric field of the laser output varies as the feedback level varies. Specifically, the sidelobes level increase or decrease, depending on the feedback level. In LADAR applications sidelobes less than −3 dB or less than half the peak value are required. In addition, a wideband signal is required in order to obtain higher resolution in the range domain. It was also observed that the bandwidth of the signal changes as the feedback level changes.

The feedback level that produces an optimal signal for LADAR applications is close to $\eta = 0.05$. The signal for this feedback level has the lower correlation entropy, a sidelobe level of 0.1828 which is less than half the maximum value, and a root mean square bandwidth of 8.38 GHz. In this case, the probability density function has a mean close to zero and is symmetric around the mean.

We also confirmed that the external cavity length does not have much influence on the resulting chaotic waveform. At the same time, the external cavity length does not have a big impact on the power spectrum of the signal. Moreover, the external cavity length has an influence on the major side lobe location on the autocorrelation. We verified that the major sidelobe location is proportional to the external cavity length. As the external cavity length is increased, the major sidelobe increases its distance from the main lobe. For that reason, an external cavity length of $L_{ext} = 0.30$ m was chosen. This yields better resolution in the range domain.

6.5 ▪ LADAR Implementation

In general LADAR and radar systems use the same principles. The difference between them is that a LADAR transmits and detects laser beams instead of RF signals as radar do. The basic operation of a LADAR is as follows, a laser is directed toward the target, some of the light is reflected back to the LADAR receiver where it is analyzed and processed to obtain the range, velocity and direction of the target. In a particular case we can discuss a Doppler radar where the LADAR is used to measure the velocity of the target. When the light beam hits a target moving towards or away from the LADAR system the frequency of the light reflected off the target will be changed creating a Doppler shift. The advantage of using a LADAR is that it gives more resolution than typical radar systems.

One of the main objectives is to obtain range and Doppler information of a target. The cross-correlation is used for obtaining the range information. Experimentally, the cross-correlation of the transmitted signal and the received signal can be obtained by using an optical interferometer. We propose to obtain the cross-correlation by using the Michelson interferometer [28].

6.5.1 ▪ Optical Correlation

The Michelson interferometer can be implemented to obtain the real-time optical cross-correlation of the transmitted and received signal of a LADAR. A schematic of the cross-correlation process is shown in Figure 6.33.

The Michelson interferometer is composed of beam splitter B, a mirror M_a, and a photodetector. The beam splitter splits the chaotic signal into two components. The stronger

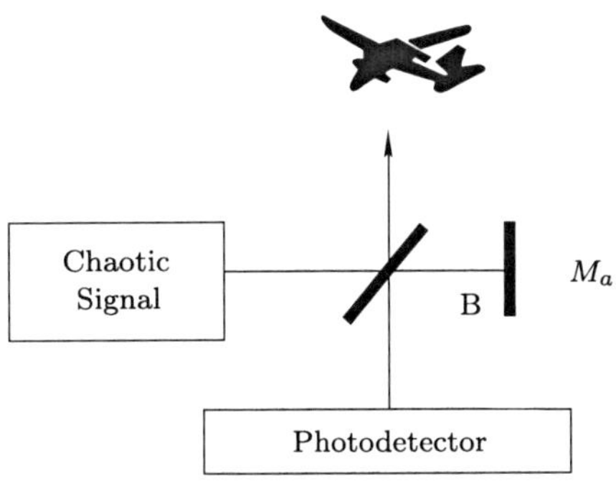

Figure 6.33. *Schematic of the cross-correlation of the chaotic field* [29].

of the beams illuminates the target, while the weaker is used for reference. The assumption is that the incident beam forms a local plane wave in the far field and illuminates the entire target. Each "hot" point of the target will contribute to the total power reflected back in the direction of the LADAR. For simplicity let the average delay be τ_0. In the case of a moving target the reflected signal has a Doppler shift $f_d = 2v/\lambda$; the reflected signal is of the form

$$r(t) = s(t - \tau_0)e^{j2\pi f_d(t-\tau_0)}. \tag{6.23}$$

Since the reference signal is reflected first by mirror M_a and then by mirror B, the photodetector responds to the sum of this signal and the echo from the target. Ideally, the photodetector's response should be a current proportional to the magnitude square of $s(t) + xr(t - \tau)$, where x is proportionality constant that depends on reflectance and transmittance of M_a, and τ is the delay associated with the target's range and the spacing between B and M_a. However, the photodetector has a finite time response that captures only slow frequencies. The net effect of this low pass filtering process is to produce a current that is proportional to the cross product $xs(t)r(t - \tau)$ integrated over a finite amount of time, giving as a result the cross correlation of the electric field:

$$R_s(\tau) = 2x \int s(t)r*(t - \tau)dt. \tag{6.24}$$

Notice that this is a scaled version of the ambiguity function of the reference signal $s(t)$. To obtain R_s as a function of the delay τ, the distance between B and M_a is changed electromechanically. Hence, we simulate the response of the interferometer as (6.2) and drop the proportionality constant.

6.5.2 ▪ Ambiguity Function

Recall that the optimized chaotic signal in the previous section was obtained for $L_{\text{ext}} = 0.30$ m and $\eta = 0.05$ by using the electric field of the laser output. These parameters are used to generate a sequence of $N = 1000$ samples. The ambiguity surface of the optimized chaotic signal is shown in Figure 6.34. Doppler steps of a fraction of the Doppler resolution are used in the simulation. On average the sidelobes in the range delay axis have a relative magnitude of $10 \log_{10}(1/N)$ with respect to the peak, which in this case is approximately −30 dB. This level is used to define the pedestal of the thumbtack ambiguity surface. The sidelobes along the range delay peak are symmetric to the origin and have an intensity of approximately −10 dB. Along the Doppler axis, the sidelobes adjacent to the main lobe have the theoretical value of −13 dB. In this case, the sidelobes in the Doppler axis is approximately −13 dB.

By using the intensity of the laser output, the ambiguity function was obtained for comparison for a feedback level of 0.05 and an external cavity length of 0.30 m. In Figure 6.35, the ambiguity function of the intensity is shown. The range delay peak is approximately −7 dB. However, a more peaked ambiguity function was obtained with the electric field. The ambiguity function of the intensity many sidelobes exist as can be observed in the next figure.

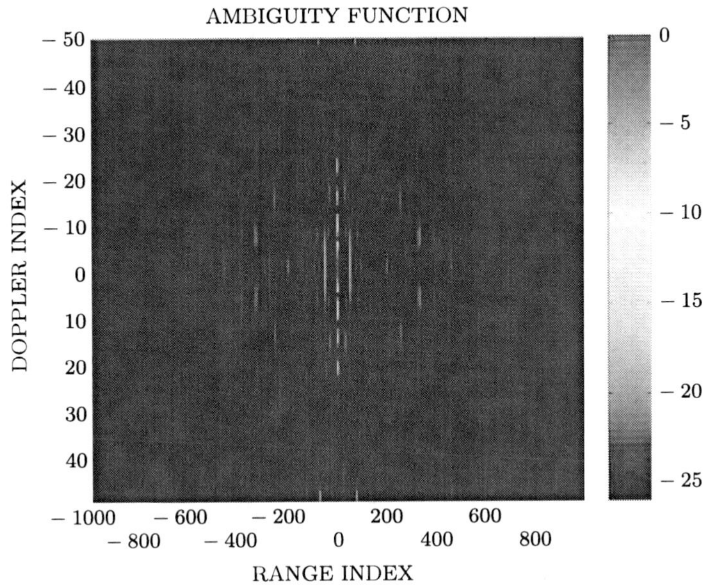

Figure 6.34. *Ambiguity function of the electric field of laser output* [29].

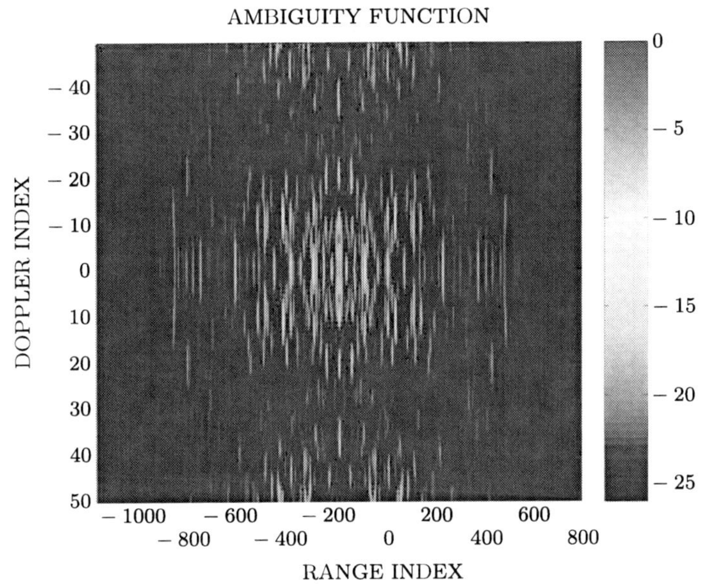

Figure 6.35. *Ambiguity function of intensity of laser output* [29].

The ambiguity function of a random field obtained from 1000 samples is shown in Figure 6.36. In this case the sidelobe levels in the range axis have an average value of —30 dB. The sidelobe levels in the Doppler axis have a level of —15 dB.

The entropy of the ambiguity function is obtained using (6.21). Elsewhere, the entropy of the ambiguity function has been used as a focal quality indicator. The image of a hot spot is perfectly focused when its entropy is a minimum. Table 6.10 shows the

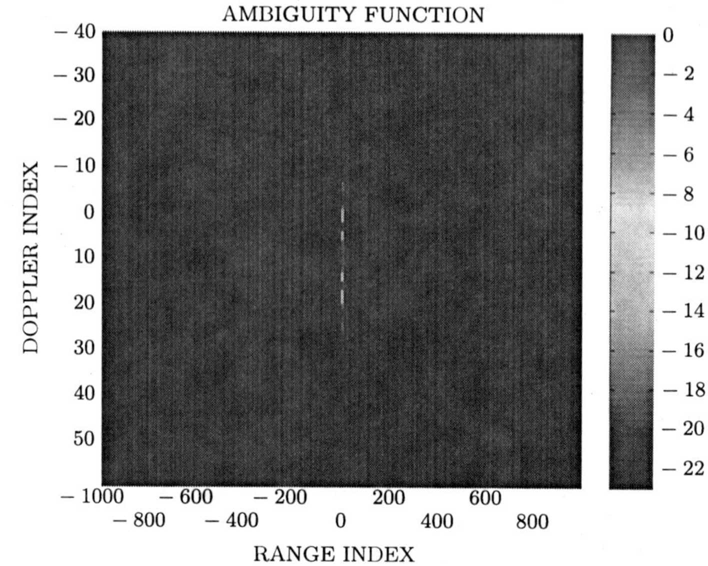

Figure 6.36. *Ambiguity function of the random signal.*

Table 6.10. *Entropy measurement for variable feedback level.*

0.02	4.964
0.03	4.923
0.04	4.933
0.05	4.897
0.06	4.929
0.07	4.908
0.08	4.905
0.09	4.901

entropy of the ambiguity function for $L_{\text{ext}} = 0.30$ m and different feedback levels. The lower ambiguity function entropy is obtained for $\eta = 0.05$. This means that the image is of higher quality when this feedback level is used.

6.5.3 ▪ hx Target Imaging

In order to simulate a more realistic target, an aircraft with multiple hot spots is chosen. The echo of the target is modeled as a coherent sum of n reflections:

$$r(t) = \sum_{i=1}^{n} s(t - \tau_i) e^{j2\pi f_{di}(t-\tau_i)}, \tag{6.25}$$

where n is the number of scatter points, τ_i is the delay associated with each scatter point, and f_{di} is the Doppler shift associated with each hot spot. For simplicity, the RADAR

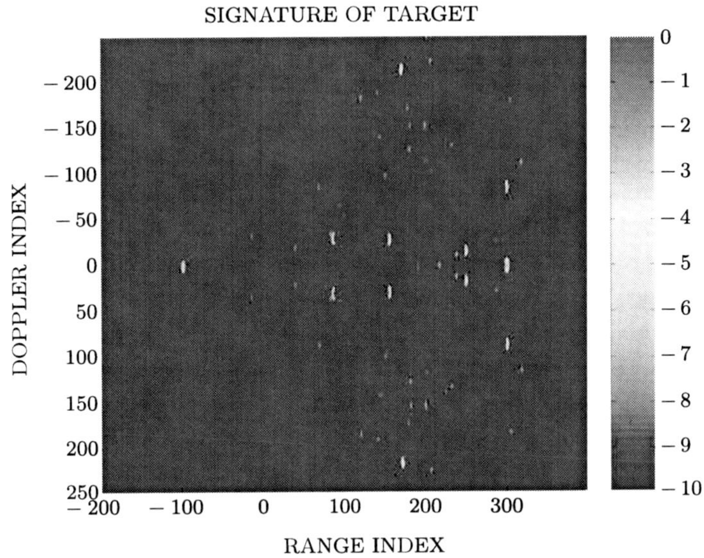

Figure 6.37. *Image of the target in the range and Doppler plane by using the electric field.*

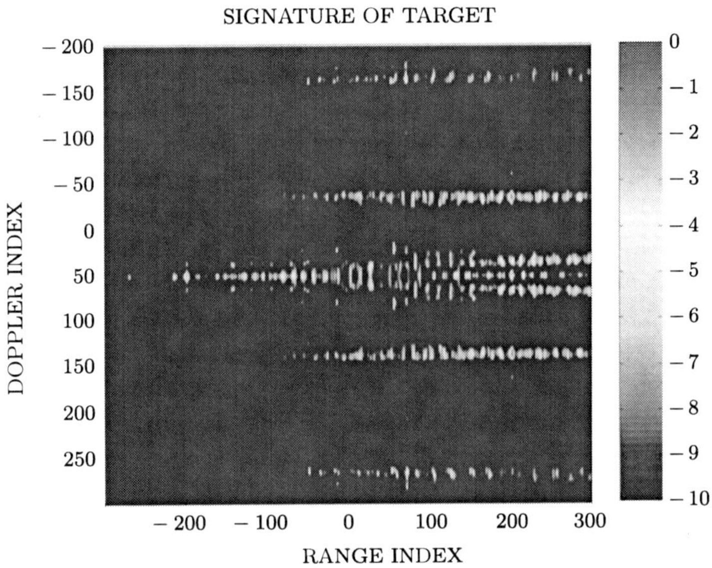

Figure 6.38. *Image of the target in the range and Doppler plane using the intensity.*

cross section of all the hot spots is assumed to be the same and independent of viewing angle. The signature of the target was obtained by using a bank of matched filters. The received signal $r(t)$ is passed through a bank of m filters. Each filter is a matched filter with a frequency f_d.

The resulting image in the range and Doppler plane is shown in Figure 6.37. In this figure, 12 hot spots were used to obtain the image of a Boeing 777. A total number of 500 samples of the electric field were used to obtain the signature of the target. For comparison, the signature of the target by using the intensity of the laser output was also obtained (Figure 6.38).

6.6 ▪ Conclusion

Chaotic dynamics are commonly modeled as complex nonlinear multivariable systems. By solving the differential equations that govern such systems, a deterministic signal can be generated with an apparent random behavior in time and frequency that depends on a set of control parameters. However, a time-frequency analysis that includes a study of the autocorrelation and spectrum of the signal is insufficient to conclude that these signals are, in the strict sense, chaotic. In this chapter, a complete characterization was performed for the electric field of a semiconductor laser modeled as a Lang–Kobayashi system and driven into chaos via optical feedback. The characterization of this three-dimensional flow required the use of pseudophase maps, attractors, and Lyapunov exponent calculations. In addition, a statistical analysis of the electric field was performed to illustrate that the electric field is indeed ergodic.

In the literature, attention is often placed on the intensity of the laser output. In doing this, the chaotic nature of the phase of the electric field is ignored. Through this chapter, it was demonstrated that the baseband representation of the electric field of the system includes both amplitude and phase information that, when used for LADAR purposes, can improve the performance of the system. This was shown by analyzing the ambiguity function of the intensity of the laser output and the electric field of the laser output as a function of time delay and Doppler shift. For an accurate description of the process, the ensemble average of the ambiguity function was obtained by introducing small changes in the initial conditions of the chaotic system. It was found that the magnitude of ambiguity function of the laser's electric field formed a nearly ideal "thumbtack" surface with sidelobes averaging a magnitude of $-20\log(N)$ dB, where N is the number of iterations of the electric field used in the computations. This result is in sharp contrast with the sidelobe level obtained by computing the ambiguity function of the laser intensity, which averaged a magnitude of $-10\log(N)$ dB. In addition, the range resolution achieved by compressing a signal proportional to the electric field was of the order of 1 mm with a compression bandwidth of the order of 10 GHz. This is an improvement of approximately 50 percent compared to the range resolution attainable by compressing a signal proportional to the intensity of the laser beam. It was also demonstrated that processing of the chaotic electric field has a real potential for use in range-Doppler imaging applications, given that the dynamic range of the responses that can be mapped spans nearly 30 dB with the processing of as few as 1000 samples per correlation. To illustrate this possibility, the signature of a Boeing 777 with 12 hot spots of identical reflectance was simulated and processed via matched filtering. The resulting imagery leaves no doubt that a clear image of the airplane can obtain by processing the electric field. The image obtained by processing the intensity of the laser is adversely affected by higher sidelobes and reduced resolution.

Acknowledgments

This work was supported by grants from the Army Research Laboratories (ARL) and the National Geospatial-Intelligence Agency (NGA).

References

[1] N. Holonyak, "Semiconductor alloy lasers-1962," *IEEE Journal of Quantum Electronics*, 23, pp. 684–691, 1987.

[2] R. Lang and K. Kobayashi, "External optical feedback effects on semiconductor injection laser properties," *IEEE Journal of Quantum Electronics*, 16, pp. 347–355, 1980.

[3] D. Lenstra et al., "Coherence collapse in single mode semiconductor lasers due to optical feedback," *IEEE Journal of Quantum Electronics*, 21, pp. 674–679, 1985.

[4] C. Masoller, "Coexistence of attractors in a laser diode with optical feedback from a laser external cavity," *Phys. Rev.*, 50, pp. 2569–2578, 1994.

[5] J. Ye et al., "Period-doubling route to chaos in a semiconductor laser with weak optical feedback", *Phys. Rev.*, 47, pp. 2249–2252, 1993.

[6] H. Li et al., "Detailed analysis of coherence collapse in semiconductor lasers," *IEEE Journal of Quantum Electronics*, 29, pp. 2421–2432, 1993.

[7] N. Kikuchi et al., "Chaos control and noise suppression in external cavity semiconductor lasers," *IEEE Journal of Quantum Electronics*, 33, pp. 56–65, 1997.

[8] J. Mork et al., "Chaos in semiconductor lasers with optical feedback: theory and experiment," *IEEE Journal of Quantum Electronics*, 28, pp. 93–108, 1992.

[9] S. Pethel and N. Corron, "Chaotic LADAR," *Applied Physics Letters*, 78, 2001.

[10] F. Y. Lin and J. M. Liu, "Chaotic lidar," *IEEE Journal of Selected Topics in Quantum Electronics*, 10, pp. 991–997, 2004

[11] J. Banks et al., *Chaos: A Mathematical Introduction*, Cambridge University Press, New York, 2003.

[12] M. Kennedy et al., *Chaotic Electronics in Telecommunications*, CRC Press, Boca Raton, 2000.

[13] B. Flores et al., "Parameter Selection of discrete chaotic maps for improved FM ambiguity surface statistics," *SIAM Conf. on Applied Dynamical Systems*, 2005.

[14] B. Flores and B. Verdin, "Wideband radar imaging using chaotic based Gaussian frequency modulation," *Proc. SPIE*, 6210, doi:10.1117/12.666300, 2006.

[15] B. Flores and G. Thomas, "Chaotic signals for wideband radar imaging," *Proc. SPIE*, 4727, doi:10.1117/12.478670, 2002.

[16] J. Sacher et al., "Intensity instabilities of semiconductor laser undercurrent modulation, external light injection and delayed feedback," *Phys. Rev.*, 45, pp. 1893–1905, 1992.

[17] T. Shuo and L. Jiaming, "Nonlinear dynamics and applications of semiconductor laser with optoelectronic feedback," *Proc. SPIE*, 4646, pp. 333–541, 2002.

[18] J. Ohtsubo, *Semiconductor Lasers: Stability, Instability and Chaos*, Springer, New York, 2005.

[19] J. Gomar, *Optical Communication Systems*, Prentice Hall, London, 1984.

[20] P. Besnard et al., "Feedback phenomena in a semiconductor laser induced by distant reflectors," *IEEE Journal of Quantum Electronics*, 29, pp. 1271–1284, 1993.

[21] W. Ray, *Stochastic and Deterministic Dynamics in a Semiconductor Laser with Optical Feedback*, Dissertation, Department of Physics, University of Maryland, 2006.

[22] S. Fukuchi et al., "Relaxation oscillation enhancement and coherence collapse in semiconductor lasers with optical feedback," *Optical Review*, 6, pp. 365–371, 1999.

[23] P. M. Alsing et al., "Lang and Kobayashi phase equation," *Phys. Rev.*, 53, pp. 4429–4434, 1996.

[24] J. C. Sprott, *Chaos and Time Series Analysis*, Oxford University Press, Oxford, UK, 2003.

[25] A. Lasota and M. Mackey, *Chaos, Fractals, and Noise: Stochastic Aspects of Dynamics*, Springer-Verlag, New York, 1994.

[26] M. Richards, *Fundamentals of Radar Signal Processing*, McGraw-Hill, New York, 2005.

[27] J. S. Son et al., *Range-Doppler Radar Imaging and Motion Compensation*, Artech House, Boston, 2001.

[28] H. Haus, *Waves and Fields in Optoelectronics*, Prentice Hall, Englewood Cliffs, New Jersey, 1984.

[29] B. C. Flores and B. Verdin, "Characterization of high resolution range and Doppler LADAR," *Proc. SPIE*, 7460, doi:10.1117/12.824712, 2009.

Chapter 7

Reverse Engineering of Complex Dynamical Systems Based on Compressive Sensing

Ying-Cheng Lai

Abstract: Recently, a paradigm of nonlinear data analysis has emerged: reconstruction of dynamical systems and complex networks based on compressive sensing. This brief review discusses the basic idea, applications, and challenges.

7.1 ▪ Introduction

The problem of predicting dynamical systems based on time series has been unresolved in nonlinear dynamics because, despite previous efforts [1] in using the standard delay-coordinate embedding method [2] to decode the topological properties of the underlying system, how to accurately infer the underlying *nonlinear system equations* remains largely an unsolved problem. In principle, a nonlinear system can be approximated by a large collection of linear equations in different regions of the phase space, which can indeed be achieved by reconstructing the Jacobian matrices on a proper grid that covers the phase-space region of interest [3, 4]. However, the accuracy and robustness of the procedure are challenging issues, including the difficulty with the required computations. In order to be able to predict system behaviors, local reconstruction of a large set of linearized dynamics is not sufficient but, rather, an accurate prediction of the underlying nonlinear equations themselves is needed.

In network science, previous efforts have been mostly on network structures and their effects on various dynamical processes taking place on or supported by the network. The types of processes that have been under intense investigations include synchronization, virus spreading, traffic flow, and cascading failures [5]. A typical approach in the field is to implement a particular dynamical process of interest on networks whose connecting topologies are completely specified. While this line of research is necessary for discovering and understanding various fundamental phenomena in complex networks, the importance of devising a general solution to the inverse problem of network prediction has been increasingly recognized [6, 7, 8, 9]. For example, in biological sciences, a significant task is to reconstruct a variety of networks from experimental data, such as inferring gene regulation networks from expression data [10, 11, 12, 13, 14]. Another example is the application of spike classification methods [15, 16, 17] to detecting interactions among neurons. Quite recently, it was demonstrated that the hierarchical property in many complex networks can be used to predict missing links [18]. Despite the success of the various existing approaches in decoding the network topology, the issue remains of whether quantitative information about node-to-node coupling, namely, the detailed dynamical coupling terms among various nodes in the network, can be inferred purely from measured time series.

Recently, a framework has been proposed [19, 20, 21] to fully reconstruct dynamical systems using time series alone, based on the assumption that the dynamics of many natural and man-made complex systems are determined by functions that can be approximated by series expansions in a suitable base. The major task is then to estimate the coefficients in the series representation. In general, the number of coefficients to be estimated can be large, but many of them are zero (the sparsity condition). According to the conventional wisdom, this would be a difficult problem, as a large amount of data is required and the computations involved can be extremely demanding. However, the recent paradigm of compressive sensing developed by Candès and others [22, 23, 24, 25, 26] provides a viable solution to the problem, where the key idea is to reconstruct a sparse signal from a small amount of observations [22, 23, 24, 25, 26], as measured by linear projections of the original signal on a few predetermined vectors. Since the requirements for the observations can be considerably relaxed as compared with those associated with conventional signal reconstruction schemes, compressive sensing has received much recent attention, and it is becoming a powerful technique to obtain a high-fidelity signal for applications where sufficient observations are not available. Here, we shall describe a general methodology to cast the problems of dynamical-system and complex-network prediction in the framework of compressive sensing.

7.2 ▪ Predicting Catastrophic Bifurcations in Nonlinear Dynamical Systems

A nonlinear dynamical system, in its parameter space, can often exhibit catastrophic bifurcations that ruin the desirable or "normal" state of operation. Consider, for example, the phenomenon of crisis [27] where, as a system parameter is changed, a chaotic attractor collides with its own basin boundary and is suddenly destroyed. After the crisis, the state of the system is completely different from that on the attractor before the crisis. Suppose that, for a nonlinear dynamical system, the state before the crisis is normal and desirable, and the state after the crisis is undesirable or destructive. The crisis can thus be regarded as a *catastrophe*, which one strives to avoid at all cost. Catastrophic events, of course, can occur in different forms in all kinds of natural and man-made systems. A question of paramount importance is how to predict catastrophes in advance of their possible occurrences. This is especially challenging when the equations of the underlying dynamical system are unknown and one must then rely on measured time series or data to predict any potential catastrophe.

Recently, we articulated a strategy [19] to address the problem of predicting catastrophes in nonlinear dynamical systems. We assume that an accurate model of the system is not available; i.e., the system equations are unknown, but the time evolutions of the key variables of the system can be accessed through monitoring or measurements. Our method consists of three steps: (i) predicting the dynamical system based on time series, (ii) identifying the parameters of the system, and (iii) performing bifurcation analysis using the predicted system equations to locate potential catastrophic events in the parameter space so as to determine the likelihood of system's drifting into a catastrophe regime. In particular, if the system operates at a parameter setting close to such a critical bifurcation, catastrophe is imminent as a small parameter change or a random perturbation can push the system beyond the bifurcation point. Once a complete set of system equations has been predicted and the parameters have been identified, one needs to examine the available parameter space. In general, to explore the multiparameter space of a dynamical system can be extremely challenging, which can often lead to the discovery of new phenomena in dynamics. An early example in this area of research is the work by Stewart et al. [28], which investigated the phenomenon of double crises in two-parameter dynamical systems. Our focus, however, was on predicting the dynamical systems based on compressive sensing.

Generally, the problem of compressive sensing can be described as the reconstruction of a sparse vector from linear measurements X about $\boldsymbol{a}$ in the form: $X = G \cdot \boldsymbol{a}$, where $X \in R^w$, G is a $w \times v$ matrix, and most components of $\boldsymbol{a}$ are zero. The compressive sensing theory ensures that the number of components of the unknown signal can be much larger than the number of required measurements for reconstruction, i.e., $v \gg w$. Accurate reconstruction can be achieved by solving the following convex optimization problem [22]: $\min \|\boldsymbol{a}\|_1$, subject to $X = G \cdot \boldsymbol{a}$, where $\|\boldsymbol{a}\|_1 = \sum_{i=1}^{v} |a_i|$ is the L_1 norm of $\boldsymbol{a}$. Solutions to the convex optimization problem have been worked out recently [22, 23, 24, 25, 26].

We argued that the inverse problem of predicting dynamical systems can be cast in the framework of compressive sensing so that optimal solutions can be obtained even when the number of base $\boldsymbol{a} \in R^v$ coefficients to be estimated is large and/or the amount of available data is small. Assume that the dynamical system can be written as $\dot{x} = \boldsymbol{F}(\boldsymbol{x})$, where $x \in R^m$ represents the set of externally accessible dynamical variables and $\boldsymbol{F}$ is a smooth vector function in R^m. The jth component of $\boldsymbol{F}(\boldsymbol{x})$ can be represented as a power series:

$$[F(\boldsymbol{x})]_j = \sum_{l_1=0}^{n} \sum_{l_2=0}^{n} \cdots \sum_{l_m=0}^{n} (a_j)_{l_1,\ldots,l_m} \cdot x_1^{l_1} x_2^{l_2} \cdots x_m^{l_m}, \tag{7.1}$$

where $x_k (k = 1, \ldots, m)$ is the kth component of the dynamical variable, and the scalar coefficient of each product term $(a_j)_{l_1,\ldots,l_m} \in R$ is to be determined from measurements. Note that the terms in (7.1) are all possible products of different components with different powers, and there are $(1+n)^m$ terms in total.

It is useful to focus on one dynamical variable of the system. (Procedures for other variables are similar.) For example, to construct the measurement vector X and the matrix G for the case of $m = 3$ (dynamical variables x, y, and z) and $n = 3$, we have the following explicit dynamical equation for the first dynamical variable: $[F(x)]_1 \equiv (a_1)_{0,0,0}x^0y^0z^0 + (a_1)_{1,0,0}x^1y^0z^0 + \cdots + (a_1)_{3,3,3}x^3y^3z^3$. We can denote the coefficients of $[F(x)]_1$ by $a_1 = [(a_1)_{0,0,0}, (a_1)_{1,0,0}, \ldots, (a_1)_{3,3,3}]^T$. Assuming that measurements of $x(t)$ at a set of time $t_1, t_2, \ldots, t_w$ are available, we denote

$$g(t) = [x(t)^0 y(t)^0 z(t)^0, x(t)^0 y(t)^0 z(t)^1, \ldots, x(t)^3 y(t)^3 z(t)^3],$$

such that $[F(x(t))]_1 = g(t) \cdot a_1$. From the expression of $[F(x)]_1$, we can choose the measurement vector as $X = [\dot{x}(t_1), \dot{x}(t_2), \ldots, \dot{x}(t_w)]^T$, which can be calculated from time series. Finally, we obtain the following equation in the form $X = G \cdot a_1$:

$$\begin{pmatrix} \dot{x}(t_1) \\ \dot{x}(t_2) \\ \vdots \\ \dot{x}(t_w) \end{pmatrix} = \begin{pmatrix} g(t_1) \\ g(t_2) \\ \vdots \\ g(t_w) \end{pmatrix} (a_1). \tag{7.2}$$

To ensure the restricted isometry property [22], we can normalize G by dividing elements in each column by the L_2 norm of that column: $(G')_{ij} = (G)_{ij}/L^2(j)$ with $L_2(j) = \sqrt{\sum_{i=1}^{M}[(G)_{ij}]^2}$, so that $X = G' \cdot a_1{}^{\cdot}$. After the normalization, $\mathbf{a}_1^{\cdot} = \mathbf{a}_1 \cdot L_2$ can be determined via some standard compressive-sensing algorithm. As a result, the coefficients a_1 are given by $a_1^{\cdot}/L_2$. To determine the set of power-series coefficients corresponding to a different dynamical variable, say y, we can simply replace the measurement vector by $X = [\dot{y}](t_1)$ and use the same matrix G. This way, all coefficients $\mathbf{a}_1$, $\mathbf{a}_2$, and $\mathbf{a}_3$ of three dimensions can be estimated.

The prediction method was tested using a number of physically relevant models of nonlinear dynamical systems in both discrete and continuous time, and was demonstrated to be able to predict catastrophic bifurcations in all cases considered [19]. A merit of this approach is that, due to the nature of the compressive-sensing method, a large number of terms can be accurately estimated even with short available time series, enabling potential implementation in real times.

7.3 ▪ Time-Series–Based Prediction of Complex Oscillator Networks

Complex networks and compressive sensing seem like two scientific topics that are far apart. Complex networks, composed of individual nodes and links among them, are mathematical and physical framework structures that distill and describe the essence and salient features of many complex interacting systems in nature and society. Various computer networks, neuronal networks, protein networks in biological cells, and collaboration networks of scientists are some of the significant and interesting examples of complex networks. As described, compressive sensing is a technique recently developed in applied mathematics and engineering signal processing to reconstruct a full picture and dynamics of the underlying system based only on a sparse amount of data. It has been widely used in all kinds of applications including, for example, the analysis of large-scale sensor networks and digital image processing in electrical engineering. It is interesting to bring compressive sensing to bear on the studies of complex networks and to address the reverse-engineering of complex networks.

The problem of uncovering the full topology of complex networks based on compressive sensing was recently addressed [20]. The basic idea is that the mathematical functions determining the dynamical couplings in a physical network can be expressed by power-series expansions. The task is then to estimate all the nonzero coefficients. Since the underlying coupling functions are unknown, the power series can contain high-order terms. The number of coefficients to be estimated can therefore be quite large. However, the number of nonzero coefficients may be only a few so that the vector of coefficients is effectively sparse, rendering applicable compressive sensing. In particular, by using power-series expansions to arbitrary order, it was demonstrated that the network-reconstruction problem can be cast in the form $X = G \cdot a$, where the vector X and matrix G are determined by the time series and a is a sparse vector to be estimated that contains all nonzero power-series coefficients in the mathematical functions of all existing couplings among the nodes. Since a is sparse, it can be solved by the standard L_1-norm technique in compressive sensing [22, 23, 24, 25, 26].

Extensive computations [20] revealed that both nonlinear nodal dynamics and node-to-node interactions can be accurately predicted, leading to reliable and robust reconstruction of the underlying networked system, as characterized by near-zero prediction errors regardless of the nature of the nodal dynamics and the network structure. Although all the examples of nodal dynamics tested are polynomial vector fields, other expansion bases were examined [20], such as trigonometric functions. If the prediction base is sufficiently wide to include all terms in the system equations as a small subset, high-accuracy prediction can be guaranteed, regardless of the mathematical forms of the terms in the equations. These features make the method appealing to predicting general complex networked systems with an extremely low data requirement.

7.4 ▪ Reconstruction of Social Networks Based on Evolutionary-Game Data

Evolutionary games are common type interactions in a variety of complex, networked natural and social systems. Given such a system, uncovering the interacting structure of the underlying network is key to understanding its collective dynamics. Recently, a general method was proposed [21] to address the problem of uncovering network topology using evolutionary-game data based on compressive sensing. Although convex optimization, the essence of compressive sensing, had been demonstrated for reconstructing coupled oscillator networks, this paradigm is powerful for solving the general inverse problem of network reconstruction based on either continuous or discrete data.

In a typical game, agents use different strategies in order to gain the maximum payoff. Generally, the strategies can be divided into two types: cooperation and defection. It was shown [21] that, even when the available information about each agent's strategy and payoff is limited, the compressive-sensing based method can yield precise knowledge of the node-to-node interaction patterns in a highly efficient manner. The method was validated by (1) extensive numerical computations using model complex networks and evolutionary games and (2) an actual social experiment in which participants forming a friendship network play a typical game to generate short sequences of strategy and payoff data. The high prediction accuracy achieved and the unique requirement of extremely small data set make the method appealing to potential applications to reveal "hidden" networks embedded in various social, economic, and biological systems.

The mathematical formulation of the method is as follows. In an evolutionary game, at any time a player can choose one of two strategies S: cooperation (C) or defection (D), which can be expressed as $S(C) = (1,0)^T$ and $S(D) = (0,1)^T$. The payoffs of two players in a game are determined by their strategies and the payoff matrix of the specific game. For example, for the prisoner's dilemma game (PDG) [29] and the snowdrift game (SG) [30], the payoff matrices are

$$P_{PDG} = \begin{pmatrix} 1 & 0 \\ b & 0 \end{pmatrix} \quad \text{and} \quad P_{SG} = \begin{pmatrix} 1 & 1-r \\ 1+r & 0 \end{pmatrix}, \tag{7.3}$$

where b $(1 < b < 2)$ and r $(0 < r < 1)$ are parameters characterizing the temptation to defect. When a defector encounters a cooperator, the defector gains payoff b in the PDG and payoff $1+r$ in the SG, but the cooperator gains the sucker payoff 0 in the PDG and payoff $1-r$ in the SG. At each time step, all agents play the game with their neighbors and gain payoffs. For agent i, the payoff is $G_i = \sum_{j\in\Gamma_i} S_i^T P S_j$, where S_i and S_j denote the strategies of agents i and j at the time and the sum is over the neighboring set Γ_i of i. After obtaining its payoff, an agent updates its strategy according to its own and neighbors' payoffs, attempting to maximize its payoff at the next round. Possible mathematical rules to capture an agent's decision making process include the best-take-over rule [29], the Fermi equation [31], and payoff-difference-determined updating probability [32]. The Fermi rule was chosen [21] in the simulations of evolutionary-game dynamics to generate a time series, which is defined as follows. After a player i randomly chooses a neighbor j, i adopts j's status S_j with the probability [31]

$$W\left(S_i \leftarrow S_j\right) = \frac{1}{1+\exp}\left[(G_i - G_j)/x\right], \tag{7.4}$$

where x characterizes the stochastic uncertainties in the game dynamics. For example, $x = 0$ corresponds to absolute rationality where the probability is zero if $G_j < G_i$ and one if $G_i < G_j$, and $x \to \infty$ corresponds to completely random decision. The probability W thus characterizes the bounded rationality of agents in society and the natural selection based on relative fitness in evolution.

The key to solving the network-reconstruction problem lies in the relationship between agents' payoffs and strategies. The interactions among agents in the network can be characterized by an $N \times N$ adjacency matrix A with elements $a_{ij} = 1$ if agents i and j are connected and $a_{ij} = 0$ otherwise. The payoff of agent x can be expressed by

$$\begin{aligned} G_x(t) =& a_{x1} S_x^T(t) \cdot P \cdot S_1(t) + \cdots \\ &+ a_{x,x-1} S_x^T(t) \cdot P \cdot S_{x-1}(t) + a_{x,x+1} S_x^T(t) \cdot P \cdot S_{x+1}(t) \\ &+ \cdots + a_{xN} S_x^T(t) \cdot P \cdot S_N(t), \end{aligned} \tag{7.5}$$

where a_{xi} $(i = 1, \dots, x-1, x+1, \dots, N)$ represents a possible connection between agent x and its neighbor i, $a_{xi} S_x^T(t) \cdot P \cdot S_i(t$ $(i = 1, \dots, x-1, x+1, \dots, N)$ stands for the possible payoff of agent x from the game with i (if there is no connection between x and i, the payoff is zero because $a_{xi} = 0$), and $t = 1, \dots, m$ is the number of round that all agents play the game with their neighbors. This relation provides us with a base to construct the vector G_x and matrix $\sum_x$ in a proper compressive-sensing framework to obtain solution of the neighboring vector A_x of agent x. In particular, we write

$$\begin{aligned} G_x &= (G_x(t_1), G_x(t_2), \ldots, G_x(t_m))^T, \\ A_x &= (a_{x1}, \ldots, a_{x,x-1}, a_{x,x+1}, \ldots, a_{xN})^T, \end{aligned} \tag{7.6}$$

and

$$\sum_x = \begin{pmatrix} F_{x1}(t_1) & \cdots & F_{x,x-1}(t_1) & F_{x,x+1}(t_1) & \cdots & F_{xN}(t_1) \\ F_{x1}(t_2) & \cdots & F_{x,x-1}(t_2) & F_{x,x+1}(t_2) & \cdots & F_{xN}(t_2) \\ \vdots & \cdots & \vdots & \vdots & \vdots & \vdots \\ F_{x1}(t_m) & \cdots & F_{x,x-1}(t_m) & F_{x,x+1}(t_m) & \cdots & F_{xN}(t_m) \end{pmatrix},$$

where $F_{xy}(t_i) = S_x^T(t_i) \cdot P \cdot S_y(t_i)$. The vectors G_x and A_x and matrix $\sum_x$ satisfy $G_x = \sum_x \cdot A_x$, where A_x is sparse due to the sparsity of the underlying complex network, making the compressive-sensing framework applicable. Since $S_x^T(t_i)$ and $S_y(t_i)$ in $F_{xy}(t_i)$ come from data and P is known, the vector G_x can be obtained directly, while the matrix $\sum_x$ can be calculated from the strategy and payoff data. The vector A_x can thus be predicted based solely on the time series. Since the self-interaction term a_{xx} is not included in the vector A_x and the self-column $[F_{xx}(t_1), \ldots, F_{xx}(t_m)]^T$ is excluded from the matrix $\sum_x$, the computation required for compressive sensing can be reduced. In a similar fashion, the neighboring vectors of all other agents can be predicted, yielding the network adjacency matrix $A = (A_1, A_2, \ldots, A_N)$.

The method was validated for model complex networks of different topologies [21]. An experiment was also conducted to reconstruct a real friendship network. In the experiment, 22 participants from Arizona State University played PDG together iteratively, and at each round each player was allowed to change his or her strategies to optimize the payoff. The payoff parameter is set (arbitrarily) to be $b = 1.2$. The player who had the highest normalized payoff (original payoff divided by the number of neighbors) summed over time was the winner and was rewarded. During the experiment, each player was allowed to communicate only with his or her direct neighbors for strategy updating. Prior to experiment, there was a social tie (link) between two players if they had already been acquainted to each other; otherwise there was no link. Each player who completed the experiment played 31 rounds of games, and he or she recorded his or her own strategy and payoff at each time, which represented the available data base for prediction. The data used for each prediction run was randomly picked from this data base. The existing friendship ties among the participants tend to favor cooperation and preclude the system from being trapped in the social dilemma because of the relatively short data streams. However, for a long run, a full defection state may occur. In this sense, the recorded data were taken during the transient dynamical phase and are thus suitable for network reconstruction. The social network was successfully uncovered, despite the complicated process of individual decision making during the experiment.

7.5 ▪ Conclusions

We have briefly reviewed the recently proposed compressive-sensing–based framework of reconstructing nonlinear dynamical systems and complex networks. The main advantage is the extremely low requirement of the amount of data. Significant challenges are that, (1) measurements of all dynamical variables are required, and (2) suitable mathematical bases are needed so that the expansions of the system equations in these bases lead to sparse coefficient vectors to be determined. Nonetheless, it is possible to apply the compressive-sensing–based method to other important contexts in science—for example, to construct gene-regulation networks from limited experimental data or digital communication networks from information flow.

Acknowledgments

The work in [19, 20, 21], on which this brief review is based, was done mainly by Dr. W.-X. Wang and Dr. R. Yang, then with Arizona State University. The author wishes to thank them. This work was supported by AFOSR grant FA9550-10-1-0083.

References

[1] H. Kantz and T. Schreiber, *Nonlinear Time Series Analysis*, Cambridge University Press, Cambridge, UK, 1997.

[2] F. Takens, "Detecting strange attractors in turbulence," in *Dynamical Systems and Turbulence, Proceedings of a Symposium held at the University of Warwick, 1979/80*, D. A. Rand and L.-S. Young (eds.), Springer-Verlag, New York, 1981.

[3] J. D. Farmer and J. J. Sidorowich, "Predicting chaotic time series," *Phys. Rev. Lett.*, 59, pp. 845–848, 1987.

[4] T. D. Sauer, "Reconstruction of shared nonlinear dynamics in a network," *Phys. Rev. Lett.*, 93, 198701, 2004.

[5] M. E. J. Newman, *Networks*, Oxford University Press, Oxford, UK, 2010.

[6] M. Timme, "Revealing network connectivity from reponse dynamics," *Phys. Rev. Lett.*, 98, 224101, 2007.

[7] D. Napoletani and T. D. Sauer, "Reconstructing the topology of sparsely connected dynamical networks," *Phys. Rev. E*, 77, 026103, 2008.

[8] W.-X. Wang, Q. Chen, L. Huang, Y.-C. Lai, and M. A. F. Harrison, "Scaling noisy fluctuations in complex networks and applications to network prediction," *Phys. Rev. E*, 80, 016116, 2009.

[9] J. Ren, W.-X. Wang, B. Li, and Y.-C. Lai, "Noise bridges dynamical correlation and topology in coupled oscillator networks," *Phys. Rev. Lett.*, 104, 058701, 2010.

[10] T. S. Gardner, D. di Bernardo, D. Lorenz, and J. J. Collins, "Inferring genetic networks and indentifying compound mode of action via expression profiling," *Science*, 301, pp. 102–105, 2003.

[11] M. Bansal, V. Belcastro, A. Ambesi-Impiombato, and D. di Bernardo, "How to infer gene networks from expression profiles," *Mol. Syst. Biol.*, 3, 78, 2007.

[12] F. Geier, J. Timmer, and C. Fleck, "Reconstructing gene-regulatory networks from time series, knock-out data, and prior knowledge," *BMC Syst. Biol.*, 1, 11, 2007.

[13] M. Hecker, S. Lambeck, S. Toepferb, E. van Someren, and R. Guthke, "Gene regulatory network inference: Data integration in dynamic models—A review," *BioSyst*, 96, 86, 2009.

[14] S. Hempel, A. Koseska, J. Kurths, and Z. Nikoloski, "Inner composition alignment for inferring directed networks from short time series," *Phys. Rev. Lett.*, 107, 054101, 2011.

[15] S. Grün, M. Diesmann, and A. Aertsen, "Unitary events in multiple single-neron spiking activity: I. Detection and significance," *Neural Comput.*, 14, pp. 43–80, 2002.

[16] R. Gütig, A. Aertsen, and S. Rotter, "Statistical significance of coincident spikes: Count-based versus rate-based statistics," *Neural Comput.*, 14, pp. 121–153, 2002.

[17] G. Pipa and S. Grün, "Non-parametric significance estimation of joint-spike events by shuffling and resampling," *Neurocomputing*, 52–54, pp. 31–37, 2003.

[18] A. Clauset, C. Moore, and M. E. J. Newman, "Hierarchal structure and the prediction of missing links in networks," *Nature*, 453, pp. 98–101, 2008.

[19] W.-X. Wang, R. Yang, Y.-C. Lai, V. Kovanis, and C. Grebogi, "Predicting catastrophes in nonlinear dynamical systems by compressive sensing," *Phys. Rev. Lett.*, 106, 154101, 2011.

[20] W.-X. Wang, R. Yang, Y.-C. Lai, V. Kovanis, and M. A. F. Harrison, "Time-series–based prediction of complex oscillator networks via compressive sensing," *Europhys. Lett.*, 94, 48006, 2011.

[21] W.-X. Wang, Y.-C. Lai, C. Grebogi, and J.-P. Ye, "Network reconstruction based on evolutionary-game data via compressive sensing," *Phys. Rev. X*, 1, 021021, 2011.

[22] E. Candès, J. Romberg, and T. Tao, "Robust uncertainty principles: Exact signal reconstruction from highly incomplete frequency information," *IEEE Trans. Inform. Theory*, 52, pp. 489–509, 2006; "Stable signal recovery from incomplete and inaccurate measurements," *Commun. Pure Appl. Math.*, 59, pp. 1207–1223, 2006.

[23] E. Candès, "Compressive sampling," *Proceedings of the International Congress of Mathematicians*, Madrid, Spain, 2006.

[24] D. Donoho, "Compressed sensing," *IEEE Trans. Inform. Theory*, 52, pp. 1289–1306, 2006.

[25] R. G. Baraniuk, "Compressive sensing [Lecture notes]," *IEEE Signal Processing Mag.*, 24, pp. 118–121, 2007.

[26] E. Candès and M. Wakin, "An introduction to compressive sampling," *IEEE Signal Processing Mag.*, 25, pp. 21–30, 2008.

[27] C. Grebogi, E. Ott, and J. A. Yorke, "Crises, sudden changes in chaotic attractors, and transient chaos," *Phys. D*, 7, pp. 181–200, 1983.

[28] H. B. Stewart, Y. Ueda, C. Grebogi, and J. A. Yorke, "Double crises in two-parameter dynamical systems," *Phys. Rev. Lett.*, 75, pp. 2478–2481, 1995.

[29] M. A. Nowak and R. M. May, "Evolutionary games and spatial chaos," *Nature*, 359, pp. 826–829, 1992.

[30] C. Hauert and M. Doebeli, “Spatial structure often inhibits the evolution of cooperation in the snowdrift game,” *Nature*, 428, pp. 643–646, 2004.

[31] G. Szabó and C. Tõke, “Evolutionary prisoner’s dilemma game on a square lattice,” *Phys. Rev. E*, 58, pp. 69–73, 1998.

[32] F. C. Santos, M. D. Santos, and J. M. Pacheco, “Social diversity promotes the emergence of cooperation in public goods games,” *Nature*, 454, pp. 213–216, 2008.

Index